Biology of Archaea 2

SCIENCES

Biology, Field Director – Marie-Christine Maurel

Microbiology, Subject Head – Marie-Christine Maurel

Biology of Archaea 2

Molecular Biology of Archaea, From Genome Maintenance to the Regulation of Gene Expression

Coordinated by
Béatrice Clouet-d'Orval
Bruno Franzetti
Philippe Oger

iSTE

WILEY

First published 2025 in Great Britain and the United States by ISTE Ltd and John Wiley & Sons, Inc.

ISTE Ltd
27-37 St George's Road
London SW19 4EU
UK

www.iste.co.uk

John Wiley & Sons, Inc.
111 River Street
Hoboken, NJ 07030
USA

www.wiley.com

Library of Congress Control Number: 2024949125

British Library Cataloguing-in-Publication Data
A CIP record for this book is available from the British Library
ISBN 978-1-78945-169-6

ERC code:
LS8 Ecology, Evolution and Environmental Biology
LS8_10 Microbial ecology and evolution

Contents

Preface

Béatrice CLOUET-D'ORVAL[1], Bruno FRANZETTI[2] and Philippe OGER[3]
[1] *MCD-CBI, CNRS, Université de Toulouse, France*
[2] *IBS, CNRS, Université Grenoble Alpes, France*
[3] *MAP, CNRS, INSA Lyon, Villeurbanne, France*

More than 40 years ago, *archaea* were proposed as a new domain of life alongside bacteria and eukaryotes. It is now accepted that these organisms constitute a distinct group, unique in many cellular and molecular aspects. The discovery of these universally important organisms represents one of the greatest advances in modern biology. Archaea have evolutionary links with early eukaryotic cells and are now used to elucidate fundamental biological questions. As champions of extremophilia, archaea have also shed light on the limits of life on Earth. Over the years, our understanding of archaeal microorganisms has evolved considerably. Far from being exotic forms of microbial life, archaea are found to be omnipresent in all terrestrial sites (>20% of marine microbial biomass) and are notably present in the human microbiota. As new genomes are sequenced, it becomes evident that archaea represent a complex, extremely diverse world and also possess a largely unexplored virosphere. In soils and oceans, archaea play a key role in the planet's major geochemical cycles. In particular, methanogens, exclusively represented by archaea, are central to issues of climate change and energy challenges. To understand their impact on the environment and human health, it is imperative to decode their unique features at a molecular level The molecular machinery deciphered in archaea will provide key paradigms for understanding fundamental biological processes conserved across life forms. Notably, the molecular bases of genetic information processing in archaea are often very similar to their counterparts in eukaryotes (translation, transcription, replication, recombination and DNA repair), even though archaea have a cellular structure similar to bacteria. Moreover, the regulatory

Biology of Archaea 2,
coordinated by Béatrice CLOUET-D'ORVAL, Bruno FRANZETTI and Philippe OGER.

systems for gene expression through post-transcriptional mechanisms remain poorly understood.

Our knowledge of archaea is rapidly evolving thanks to the advent of high-throughput sequencing of entire genomes. It is time to introduce a wide audience to the advances that illuminate our understanding of biology and reveal the originality of this domain of life.

The 14 chapters, divided into two volumes, review the discovery and evolution of the domain of archaea and summarize our current knowledge of cellular and molecular processes to better integrate this domain into the broader understanding of life.

January 2025

1

Replication of Archaeal Chromosomes

Ghislaine HENNEKE[1], Roxane LESTINI[2], Marc NADAL[3] and Didier FLAMENT[1]

[1]*BEEP, CNRS, Ifremer, Université de Bretagne Occidentale, Plouzané, France*

[2]*Laboratoire d'optique et biosciences, Inserm, CNRS, Institut Polytechnique de Paris, Palaiseau, France*

[3]*IBENS, CNRS, Inserm, PSL University, Paris, France*

It was only when Watson and Crick proposed the double helix structure of DNA, in the early 1950s, that the scientific community finally accepted that DNA could be the carrier of heredity. Since then, the study of DNA replication has become one of the key themes of molecular biology. First developed in bacteria in the 1960s, then in eukaryotes in the 1980s, it was only after Carl Woese and George Fox clarified the phylogenetic position of archaea that the study of the molecular mechanisms of DNA replication began in archaea, in the early 1980s. This new scientific venture has proved exciting in several ways. First, although the composition of the replication machinery shows a high degree of diversity in archaea, it is similar to that of eukaryotes, indicating a common evolutionary history (see Volume 1, Chapters 2 and 3).

Thus, archaeal proteins and the metabolic pathways involved in this process represent excellent models for study. Furthermore, the study of replication initiation in these organisms has recently challenged one of the dogmas of molecular biology, according to which an origin of replication is required to initiate the autonomous

Biology of Archaea 2,
coordinated by Béatrice CLOUET-D'ORVAL, Bruno FRANZETTI and Philippe OGER.

replication of a DNA fragment. Indeed, in some archaea, the origin(s) of replication (of which there can be several) can be genetically suppressed without the cell exhibiting growth defects. This suggests the presence of an alternative replication initiation mechanism in archaea, which remains to be characterized.

The general process of DNA replication is conserved throughout all cellular organisms. DNA replication can be divided into four main stages. The first stage involves initiation, during which specific proteins recognize and bind to the origin of replication. This DNA–protein complex recruits a helicase to unwind the double-stranded DNA at the origin. DNA primase and DNA polymerases are then recruited to the replication bubble generated by the helicase activity. The second stage is elongation, which corresponds to the duplication of both DNA strands simultaneously and bidirectionally. Due to the antiparallel structure of DNA and the direction of nucleotide polymerization from 5′ to 3′, the leading strand is synthesized continuously, while the lagging strand is replicated discontinuously in the form of Okazaki fragments. In the third stage, the RNA primers on the lagging strand are removed by the coordinated action of ribonucleases, flap endonucleases and DNA ligases. Finally, the last stage, known as termination, corresponds to the end of synthesis, when the replication forks meet or when they encounter a termination signal.

In this chapter, we will summarize what we know about DNA replication in archaea through the different phases of this process.

1.1. Replication initiation

1.1.1. *Identifying the origins of replication*

Although we now know the complete genome of some 420 cultivated isolates, our knowledge of the replication of these organisms is limited to <20 species, some of which are very closely related, and do not cover all the phyla of the archaea domain. In terms of genome structure, very different situations coexist, since some archaea have a single circular chromosome, like most bacteria, while halophilic archaea (*Haloarchaea*) have several chromosomes. Furthermore, while some archaea are haploid, others are polyploid. This genomic diversity implies different couplings with the cell cycle and different regulations of replication initiation.

The first origin of replication observed in archaea was that of *Pyrococcus abyssi*. This *Euryarcheota*, of the order *Thermococcales*, has a single chromosome containing a single bidirectional origin of replication (Matsunaga et al. 2001). In methanogens, in silico studies indicate the presence of a single origin. Allele frequency analysis in *Nitrosopumilus maritimus*, a *Thaumarchaeota*, reveals the presence of a single origin of replication. However, multiple origins of replication per chromosome have been found in *Halobacteria*. In *Crenarchaeota*, there seems to be the presence, on a single circular chromosome, of up to four multiple origins of replication, this being unrelated to genome size. Surprisingly, in 2013, Thorsten Allers' team at the University of Nottingham demonstrated that all the origins of replication in the halophilic archaea *Haloferax volcanii* could be deleted without preventing the mutant strain from dividing and continuing to grow (Hawkins et al. 2013). This same finding was then established in a hyperthermophilic archaea of the order *Thermococcales* (Gehring et al. 2017), raising the question of the role of these origins of replication in certain archaea, as well as the alternative mechanism enabling chromosome replication in the absence of the origin of replication. Although these studies still only concern a small number of archaea, it seems that the vast majority of archaea have multiple origins of replication, with *Thermococcales* and *Thaumarchaeota* being the exceptions. To date, all observed origins of replication allow bidirectional replication of the associated replicon.

1.1.2. *Recognizing the origins of replication*

Replication initiation takes place at specific sites, composed of at least two short repetitive sequences, the ORBs (origin recognition boxes) (Matsunaga et al. 2001; Robinson et al. 2007), located near an AT-rich DNA sequence enabling local separation of the two complementary strands, known as DNA unwinding element (DUE) (Figure 1.1(a)). These sequences are recognized by specific proteins, the Orc/Cdc6 proteins. These proteins belong to the same family as the Orc (origin recognition complex) proteins involved in the initiation of replication in eukaryotes. Their binding induces local DNA unwinding (Matsunaga et al. 2009). In eukaryotes, the six Orc proteins (Orc1–6) associate at replication origins with Cdc6 and Cdt1 proteins, so as to enable initiation. In archaea, the initiator proteins are both homologous to the Orc protein family, essentially Orc1, and to Cdc6 proteins. It was surprising to find these proteins in multiple copies, with more than 14 in *H. volcanii* (Norais et al. 2007). Furthermore, genes are most often localized in the immediate vicinity of the origin to which they bind, reflecting the localization of the bacterial initiator protein DnaA (Figure 1.1). *Sulfolobus islandicus* has been shown to possess three replication origins, OriC1, OriC2 and OriC3, which bind Orc1-1, Orc1-3 and WhiP proteins, respectively. Unlike the other initiation proteins, WhiP is related to

Cdt1, a protein involved in origin recognition in eukaryotes. Although the order *Sulfolobales* has an Orc1-2 protein, it is WhiP that binds to the OriC3 region.

As in eukaryotes, expression of the genes encoding Orc/Cdc6 proteins is finely regulated during the cell cycle. In *Sulfolobales*, genes encoding Orc1-1 and Orc1-3 proteins are expressed at the start of the G1 phase (Lundgren and Bernander 2007), leading to the binding of these initiator proteins to replication origin sequences at the start of the G1 phase. The Orc1/Cdc6 proteins have an ATP binding and hydrolysis site, and conformational changes take place during the ATP binding/hydrolysis cycle. ATP binding is necessary for the activation of replication origins, but ATP hydrolysis is not essential for replication. For the moment, the precise signals leading to replication induction are not known. However, the third Orc1/Cdc6-like protein, Orc1-2, which is not involved in replication initiation, could act as a replication repressor. The sequence recognized by Orc1-2 overlaps with those recognized by Orc1-1 and Orc1-3 (at OriC1 and OriC2 origins, respectively). Consequently, its binding would competitively block these two origins, which is consistent with the fact that this protein is specifically expressed in the G2 phase.

While the main mode of replication in archaea seems to be the replicon hypothesis proposed by Jacob et al. (1963) – in other words, involving the attachment of an inducible protein transacting on a particular sequence – the presence of these origins, and the proteins capable of recognizing them, is not essential in archaea. In fact, in *H. volcanii*, deletion of the four replication origins of the largest chromosome enables faster growth of these cells without it being possible to identify a dormant origin, as is the case in *Haloferax mediterranei* (Yang et al. 2015). In the *H. volcanii* mutant without the origin sequences, the presence of the RadA protein, the homologue of the RecA/Rad51 homologous recombination proteins, is required to allow for replication (Hawkins et al. 2013). The same is true for *Thermococcus kodakarensis*, in which deletion of the single origin of replication and/or the Cdc6 protein does not alter the growth of this archaea.

This type of replication, known as RDR (recombination-dependent DNA replication), makes use of homologous recombination, as already described in bacteria (Ogawa et al. 1984), but with the consequence of reduced growth in bacteria. The precise mechanisms leading to the formation of a complex-enabling genome replication through this pathway could be involved in the polyploidy observed, but they are, as yet, unknown to us, and we will therefore concentrate hereafter on replication, starting from the sequences of replication origin.

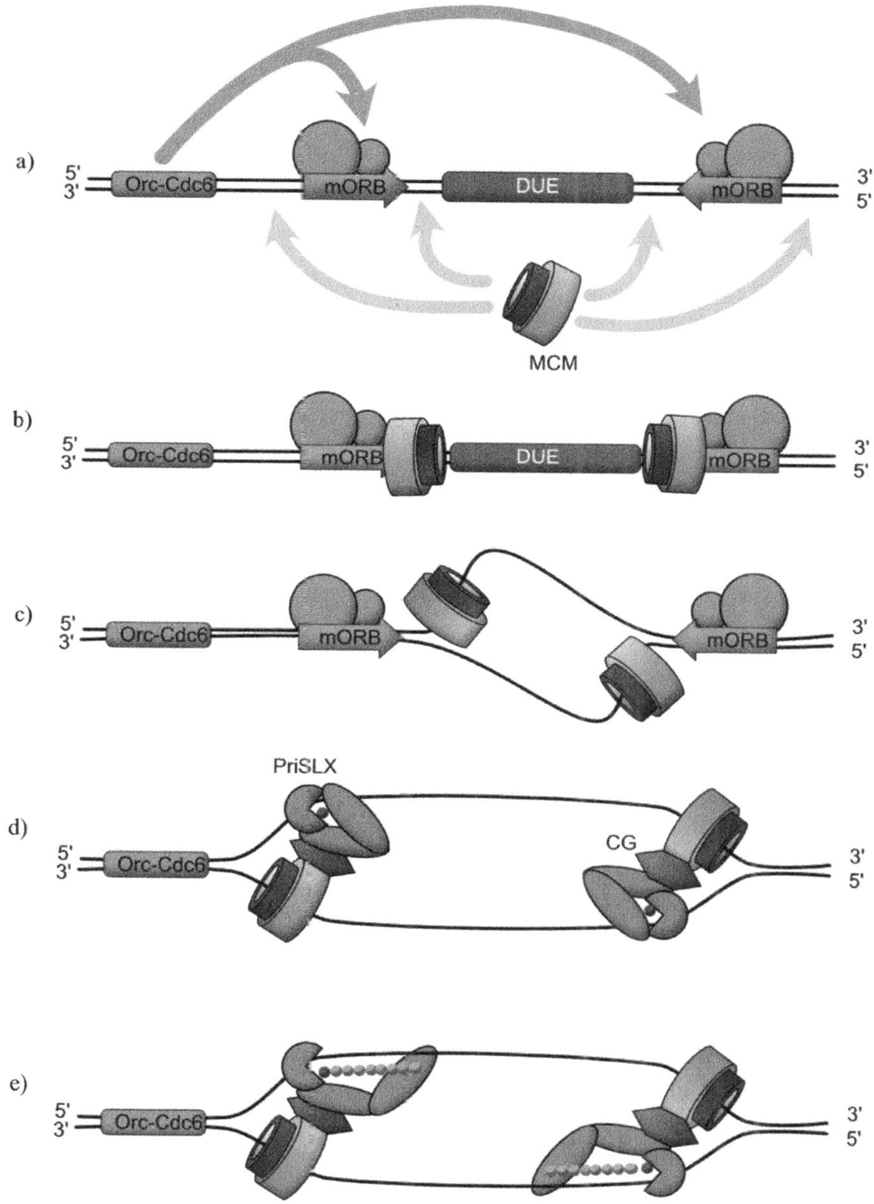

Figure 1.1. *The different phases of replication initiation . For a color version of this figure, see www.iste.co.uk/clouet/archaea2.zip*

COMMENT ON FIGURE 1.1.– *(a) A common feature of the archaeal origins of replication is the presence of a DUE (DNA unwinding element) sequence flanked by adjacent Orc-cdc6 (mORB) protein binding sites. The gene specifying the Orc-cdc6 initiator protein is found close to the origin. The Orc-cdc6 initiator protein binds to the mORB sites and then facilitates the recruitment of MCM replicative helicases; (b) by interacting preferentially with the open forms of MCM, Orc-cdc6 would thus enable their positions, which would probably take place on double-stranded DNA; (c) similar to the situation in eukaryotes, a transition would enable MCM to bind to single-stranded DNA in a targeted way and would cause the opening of the DUE region; (d) the CG complex ((Gins14-Gin23-Cdc45)2) binds to and activates the MCM helicase, enabling it to move by hydrolyzing ATP, which leads to dissociation of the Orc-cdc6 proteins, thus allowing primase binding; (e) primase (PriSLX) then binds the first nucleotide triphosphate (in red), and then synthesizes a short RNA chain forming the primer.*

1.1.3. *Pre-initiation complex formation*

As in eukaryotes, it is the ORB-Orc1/Cdc6 complex that serves as the platform for the replisome. A critical step is the positioning of the replicative helicases. In archaea, the replicative helicase belongs to the MCM family of proteins, but unlike in eukaryotes, where it forms a ring consisting of six different subunits (MCM2-7), it is a homohexamer in archaea. Two helicases are positioned head-to-tail on the DNA at the ORB-Orc1/Cdc6 complex, each hexamer surrounding a double-stranded DNA (Figure 1.1(b)). Archaeal MCMs naturally exist in two forms, closed and open. The open form of the ring allows the double-stranded DNA to be positioned centrally, without the need for a helicase loader, unlike in bacteria and eukaryotes. The Orc1/Cdc6 proteins have two roles. On the one hand, they enable correct positioning of the two MCMs at the region of origin, possibly by facilitating the opening of the MCM hexamer; on the other hand, they generate a single-stranded DNA region at the DUE region. A rearrangement of the complex (that is yet-to-be-understood) would allow the positioning of a single-stranded DNA in the center of the ring. It is in this form that MCMs acquire their helicase activities, moving onto this single strand and ejecting the other strand of the DNA duplex (Figure 1.1(c)). These helicases become fully active after association with the (Gins23-Gins15)2 complex, forming a homolog of the eukaryotic GINS complex and two copies of the eukaryotic Cdc45 homologous protein. The GINS-MCM-Cdc45 combination then reconstitutes a GINS-Cdc45 complex that is capable of strongly stimulating MCM helicase activity, forming a GMC complex similar to that of eukaryotes (Greci and Bell 2020) (Figure 1.1(d)).

We have seen that a single strand of DNA is present in the central channel of the MCM hexamer, enabling it to move from 3′ to 5′ on the single-stranded DNA serving as a template for continuous replication. To enable this advance and the opening of the DNA, the helicase hydrolyzes ATP. The transition that allows complete opening of the original region and correct positioning of MCM helicases on single-chain DNA has yet to be determined. To form a bidirectional replication bubble, it has been proposed that the two helicases cross over one another, passing one above the other, each on its own strand. Single-stranded DNA-binding proteins would then act, essentially on the discontinuous strand, with the primase on the continuous strand. Archaea possess two types of single-stranded DNA-binding proteins, one very similar to the RP-A protein of eukaryotes, the other very similar to the SSB protein of bacteria. Primase (PriSLX) is then likely to be positioned simultaneously, as primase interacts with the GMC complex via the GINS complex (Figure 1.1(d)). Presumably, this interaction is what ensures the coupling between the progression of the replication fork, created by the helicase and primer synthesis (Greci and Bell 2020) (Figure 1.1(e)). At this stage, activation of the GMC complex leads to the formation of a replication bubble, on which other replisome components are recruited to establish two replication forks and initiate bidirectional DNA synthesis.

1.1.4. *Primer synthesis*

DNA primases play an essential role in initiating DNA replication by synthesizing RNA primers on both the leading and lagging strands. In fact, DNA polymerases are unable to synthesize DNA de novo, and require a primer with a 3′-OH end to elongate DNA chains. In contrast, DNA primases are able to synthesize de novo short RNA fragments on template DNA, which serve as primers and are extended by the DNA polymerase (Figure 1.1(e)). Archaeal DNA primases consist of two subunits, PriS and PriL, which are conserved throughout the archaeal domain. However, a third component of the primase complex, PriX, is also present in many representatives of the *Crenarchaeota* phylum. PriS contains the enzyme's active site, and PriL or PriLX regulates primase activity and contains an iron–sulfur cluster. In addition, PriS and PriL have orthologs in eukaryotes that form the primase module of the DNA polymerase α-primase complex. The PriSL primase complex is thus sometimes referred to as archaeal eukaryotic primase. Both subunits are essential for cell viability. However, the archaeal primase has the unique feature of being able to synthesize both DNA and RNA in vitro. Indeed, the PriS subunit of *Pyrococcus furiosus* synthesizes long DNA fragments, but the addition of the PriL subunit regulates the process by reducing DNA polymerase activity and increasing RNA polymerase activity, while decreasing the length of the fragments being generated (Liu et al. 2015). Furthermore, studies on the *Pyrococcus abyssi* enzyme suggest that the primase may also be involved in DNA repair, due to its DNA polymerase nucleotide gap filling and strand displacement activities, also observed in vitro (Le Breton et al. 2007). Although archaeal DNA primases exhibit both DNA and RNA synthesis activities in vitro, it has been shown that the primase is predominantly an RNA synthesis enzyme when the ratio of ribonucleotides (NTPs) to deoxyribonucleotides (dNTPs) is close to physiological values (Yan et al. 2018). Thus, the most likely hypothesis is that archaeal DNA primase is able to synthesize a short RNA fragment in the replication bubble on the forward strand, then steadily on the lagging strand, and that a primase conformational change mechanism governs the length of the synthesized fragments of 12–14 nucleotides (Greci and Bell 2020) (Figure 1.1(e)).

1.2. The elongation phase of replication

Following primer synthesis, the RNA primer synthesized by the primase is extended by a so-called "replicative" DNA polymerase. DNA polymerases have been classified into seven families on the basis of amino acid sequence similarity (A, B, C, D, E, X and Y). Bacteria and eukaryotes have five and 15 DNA polymerases, respectively. Replicative polymerases belong to the C family in bacteria and to the B family in eukaryotes. In archaea, the diversity of DNA polymerases is less rich, given that depending on the species, only two to three polymerase families are represented, by two to four different polymerases. These are the B, D and Y families. The DNA polymerases of the Y family are involved in DNA repair mechanisms, while the B and D families contain replicative DNA polymerases, which have a different phyletic distribution within the archaea domain. Archaeal DNA polymerase D, PolD, was the last DNA polymerase to be discovered, in the late 1990s (Cann et al. 1998). It is specific to archaea and is present in all phyla, except the *Crenarchaeota*. PolD is an enzyme complex composed of two subunits: DP1, which carries the proofreading activity, and DP2, which carries the DNA polymerization activity. The recent acquisition of the crystallographic structure of PolD has shown that the structure of DP2 resembles the catalytic center of RNA polymerases, thus opening a window into the evolutionary transition between the RNA and DNA worlds (Sauguet 2016). In terms of function, PolD is essential for viability, suggesting that it is the replicative DNA polymerase in species that possess it. Indeed, these species also generally possess another B-family polymerase, PolB, which is not essential for viability and is more likely to be involved in DNA repair activities (Kushida et al. 2019). *Crenarchaeota*, which do not possess PolD, have at least two B-family DNA polymerases that ensure genome replication (PolB I and PolB II, sometimes accompanied by PolB III), one of which is essential, namely PolB I.

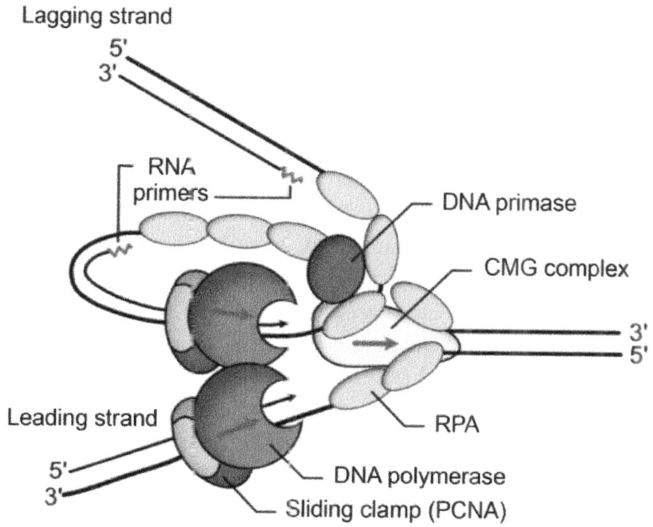

Figure 1.2. *Diagram of the composition and organization of the replisome. For a color version of this figure, see www.iste.co.uk/clouet/archaea2.zip*

As with all replicative DNA polymerases, PolD or PolB possess both DNA synthesis and error correction activities. These capabilities enable them to accurately replicate the genome, generating few errors. However, while this characteristic is essential for chromosome replication, it is not sufficient to ensure the completion of this cellular process. In particular, replicative DNA polymerases are always associated with ancillary factors that give them qualities of processivity (average number of nucleotides added to the DNA strand being synthesized, without the polymerase detaching itself from the template strand; this can exceed 100,000 nucleotides for replicative polymerases) and speed. This is particularly true of one of the essential factors in the architecture of the replisome for all three branches of life, the so-called "sliding clamp" protein (Figure 1.2). This sliding clamp surrounds the DNA and binds to the DNA polymerases, enabling the enzyme to remain in contact with the template DNA strand, thus improving the enzyme's processivity. This factor (PCNA in eukaryotes and archaea, or *β-clamp* in bacteria) has a ring-like structure, formed by the assembly of two (*β-clamp*) or three (PCNA) subunits, and functions as a molecular platform that is capable of binding to DNA polymerases, as well as many other DNA-modifying proteins. To load the clamp onto the double-stranded DNA, a second factor is required: replication factor C (RFC).

Its role is to attach the sliding clamp and load it onto the DNA at the 3′-OH end of the primer (Kelman and Kelman 2014). The organization of these factors, such as replicative DNA polymerase, PCNA, CMG complex and DNA primase, that make up the replisome allows the coupling of the replication fork progression and DNA synthesis. As the replication fork advances, it causes a separation of the two upstream DNA strands, generating two single-stranded DNAs that serve as templates for DNA synthesis (Figure 1.2). These single-stranded portions of DNA are highly unstable and prone to degradation. In order to prevent such damage, the single-stranded portions of DNA at the fork are colonized by a single-stranded DNA-binding protein (RPA for *Euryarchaeota* and SSB for *Crenarchaeota*), which prevents their degradation by nucleases (Taib et al. 2021). The spatial and time-based organization of these different components (DNA polymerases, primase, CMG complex, PCNA, RFC and RPA) at the replication fork level is not yet well defined in archaea (Figure 1.2). However, this replication fork will enable a coordinated DNA synthesis on the two strands, even though they have opposite polarities. Indeed, the anti-parallel nature of the DNA double helix and the 5′→3′ directionality of replicative DNA polymerases dictate that the leading strand, which has been newly synthesized at the replication fork, is polymerized continuously, while the other is replicated discontinuously. In other words, the primase regularly synthesizes RNA primers on the lagging strand, which are then elongated by the DNA polymerase, generating a succession of multiple segments known as Okazaki fragments (Okazaki et al. 1968) (Figure 1.2), whose maturation will generate a continuous strand of neosynthesized DNA.

1.3. Okazaki fragment maturation

The initiator RNA at the 5′ end of the Okazaki fragments must be removed and replaced by DNA. These newly synthesized DNA fragments are then joined together, creating a fully replicated, continuous DNA. This process is called Okazaki fragment maturation. It involves the coordinated action of various proteins and enzymatic activities. In particular, RPA (replication protein A), PCNA (proliferating cell nuclear antigen), a DNA polymerase, two nucleases FEN1 (Flap endonuclease 1) and RNase H (ribonuclease H), and a DNA ligase are involved in this mechanism. The essential steps in Okazaki fragment maturation are based on (i) ribonucleotide removal by ribonuclease H type 2 (RNase HII); (ii) oligonucleotide removal by FEN1; and (iii) DNA fragment assembly by DNA ligase. This process mainly takes place on the discontinuous strand, but also acts at the origin level, in the region of divergences between the continuous and discontinuous strands.

1.3.1. *Removal of ribonucleotides by RNase HII*

RNase HII is an enzyme capable of recognizing and hydrolyzing ribonucleotides incorporated into double-stranded DNA. It specifically cuts the phosphodiester bonds of the RNA moiety, generating a DNA fragment containing a 5′-mono-ribonucleotide (Figure 1.3(a) II). Unlike ribonuclease H type 1 or RNase HI, which cleaves RNA–DNA hybrids containing at least four RNA residues, RNase HII can catalyze the 5′ cleavage of a single ribonucleotide. These various cutting specificities suggest the involvement of ribonucleases H in mechanisms that maintain genomic stability (Tadokoro and Kanaya 2009). With the exception of a few archaea that have both types of ribonuclease H (RNases HI and HII), such as *Sulfurispaera tokodaii*, *H. volcanii*, *Halobacterium* sp. *NRC-1* or *Pyrobaculum aerophilum*, the majority of archaeal genomes contain a single ribonuclease H, that being RNase HII. Thus, the ubiquity of this enzyme suggests its functional multidisciplinarity in DNA repair or replication pathways in archaea, even though it does not seem essential for cell survival (Meslet-Cladiére et al. 2007; Sarmiento et al. 2013; Birien 2018). Moreover, PCNA, an essential protein factor in genomic maintenance processes, is able to modulate the enzymatic properties of RNase HII (Meslet-Cladiére et al. 2007).

Solved three-dimensional (3D) structure of RNases HII is available in various archaea (Lai et al. 2000). It contains a highly conserved catalytic core, a characteristic of all ribonuclease 3D structures (Clouet-d'Orval et al. 2018). Like most other replication proteins, the structure–function relationship of archaeal RNase HII shares more homologies with eukaryotes than with bacteria (Malfatti et al. 2019). The mechanism for excising ribonucleotides from DNA is the most representative cellular function of RNase HII in the three domains of life, with the enzyme cutting 5′ to the ribonucleotide, initiating the repair system (Heider et al. 2017). Although the enzyme is not essential in archaea, as it is in yeast (Jeong et al. 2004), it cannot be ruled out that RNase HII may be involved as a backup or in cooperation in the Okazaki fragment maturation process (Henneke 2012) (Figure 1.3(a)). In this case, the RNase HII cleavage product would initiate the Okazaki fragment maturation pathway, preceding the strand synthesis-displacement step by DNA polymerase, followed by FEN1 cleavage of the branched double-stranded DNA (Figure 1.3(a) II).

1.3.2. *Removal of oligonucleotides by FEN1*

FEN1 is a structure-specific nuclease for DNA and/or RNA. It has endonuclease and 5′–3′ exonuclease activity (Friedrich-Heineken et al. 2003; Henneke et al. 2003). Like its eukaryotic counterpart, FEN1 plays a significant role in Okazaki fragment maturation and DNA repair (Henneke et al. 2003; Balakrishnan and Bambara 2013; Burkhart et al. 2017). It interacts directly and is stimulated by PCNA (Henneke 2012; Heider et al. 2017). In the presence of RNase HII (Figure 1.3(a)), the Okazaki fragment maturation process requires the removal of the mono-ribonucleotide from the preceding fragment in order for ligation by DNA ligase to proceed correctly. In this pathway, the replicating DNA polymerase (family B or D) during DNA synthesis encounters and displaces the fragment containing the mono-ribonucleotide, generating a branched double-stranded DNA substrate recognized and cut by FEN1 (Sato et al. 2003; Henneke 2012). In the absence of RNase HII (Figure 1.3(b)), complete removal of the initiator RNA from Okazaki fragments relies on the coordinated action of synthesis and strand displacement by the replicative DNA polymerase (B or D family), as well as FEN1.

The length of the displaced single strand is maintained at a minimal size, so as to avoid exposure to various rearrangements or degradation (Henneke 2012). Occasionally, strand-displacement activity by DNA polymerase can become ineffective. In this case, degradation of the DNA-hybridized initiator RNA would rely on activation of the 5′–3′ exonuclease function of FEN1. Although not reported in archaea, this maturation mechanism could be similar to the nick translation process encountered in eukaryotes and bacteria (Lundquist and Olivera 1982; Stodola and Burgers 2016).

The three-dimensional structure of FEN1, alone or in the presence of its substrate, has been obtained in various archaea (Hosfield et al. 1998; Chapados et al. 2004). The catalytic domain of FEN1 resembles that of nucleases and contains an active site with two metal ions. Catalysis relies on a conformational change in the FEN1 helical loop which, after binding to the intermediate structures of the Okazaki fragments, lead to the 5′ cleavage specificity of the fragment (Storici et al. 2002). It would appear that the archaeal FEN1 protein is capable of recognizing, binding and cutting a branched double-stranded fragment with a 1-nucleotide unpaired 3′ end (downstream strand) and a short flap containing the initiator 5′ RNA (upstream strand) (Chapados et al. 2004). Given the importance of FEN1 in the maturation process of Okazaki fragments, it is however surprising that its genetic deletion in archaea (Meslet-Cladiére et al. 2007) does not result in cell death, with the exception of *Halobacterium sp. NRC-1* (Berquist et al. 2007). Like in yeast (Storici et al. 2002), it is highly probable that archaea use substitution nucleases. For example, the exonuclease activity of the GINS-associated nuclease (GAN) complex in *T. kodakarensis* has been suggested to replace the combined action of the two nucleases RNase HII and FEN1 (Burkhart et al. 2017).

1.3.3. *DNA fragment assembly by DNA ligase*

After complete removal of the initiator RNA from the Okazaki fragments, DNA ligase (LIG1) sutures the newly synthesized DNA strand, replacing the RNA and adjacent DNA fragment (Figure 1.3(a) V and (b) IV).

Catalysis involves the formation of a phosphodiester bond between the 3′-OH carbon of the upstream strand and the 5′-phosphate of the adjacent strand. It involves a cofactor as an energy source, namely, ATP and/or NAD+, the use of which by LIG1 is dependent on the type of archaea (Rolland et al. 2004; Zhao et al. 2006). DNA ligase interacts physically with PCNA, which stabilizes it on the DNA duplex (Vijayakumar et al. 2007).

With the exception of archaea that possess heterotrimeric PCNA, such as *Saccharolobus solfataricus*, it has been suggested that DNA polymerase, FEN1 and DNA ligase compete for the same homotrimeric PCNA-binding sites (Henneke 2012), as described in eukaryotes (Ayyagari et al. 2003).

The three-dimensional structure of LIG1 has been solved in various archaea (Pascal et al. 2006), and the architectural organization of its three domains (the nucleotidyl transferase domain (NT), the oligonucleotide/oligosaccharide folding domain (OB-fold) and the DNA-binding domain (DBD)) strongly resembles that of human LIG1 in complex with DNA (Pascal et al. 2004). Although the conformational dynamics of these three domains differ between archaeal and eukaryotic LIG1s, they are involved in the implementation of catalysis.

In the case of ATP-dependent LIG1, the OB domain transfers AMP to a lysine in the active site of the NT domain during the first step of the reaction (Sriskanda and Shuman 1998). AMP is then transferred from LIG1 to the DNA 5′-phosphate group, and LIG1 catalyzes the final ligation step, releasing the AMP.

The DBD domain, which is largely responsible for the physical interaction of DNA ligase with its substrate, enables the three domains (DBD-NT-OB) to completely envelop and accurately position the NT and OB domains involved in catalysis at the 3′-OH and 5′-phosphate ends of the DNA duplex (Pascal et al. 2004).

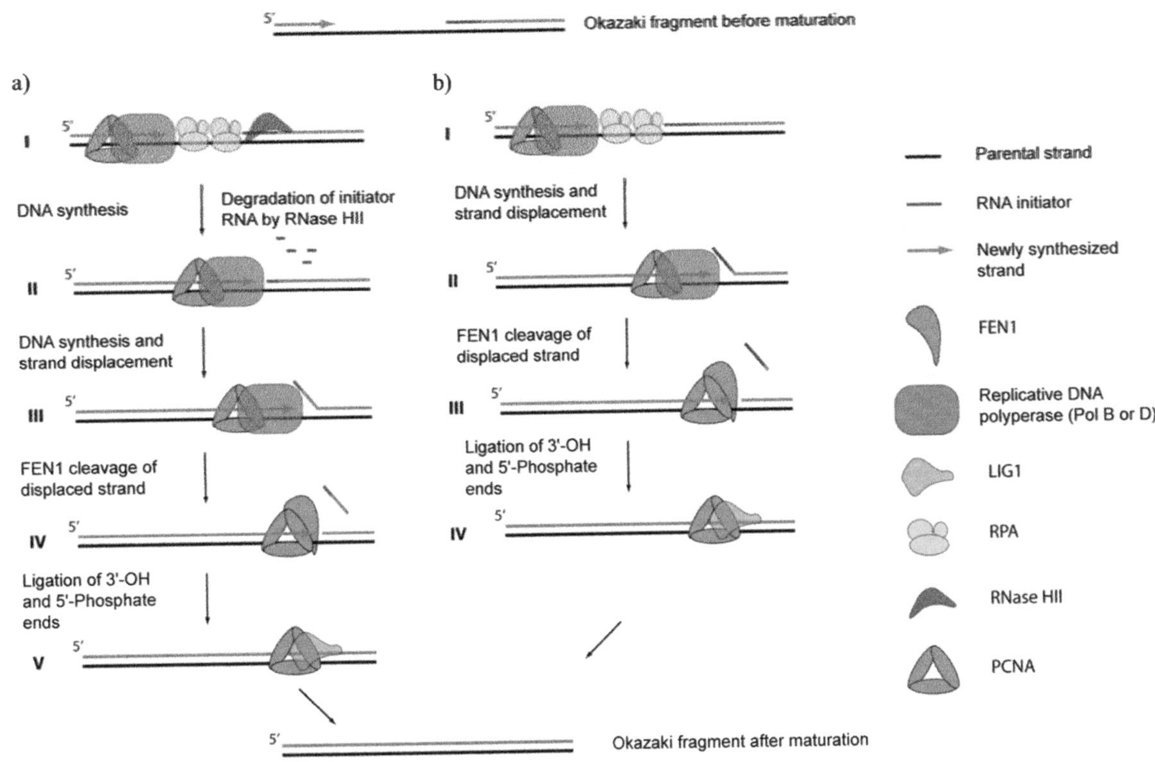

Figure 1.3. *Maturation mechanism of archaeal Okazaki fragments. For a color version of this figure, see www.iste.co.uk/clouet/archaea2.zip*

COMMENT ON FIGURE 1.3.– *(a) Reaction pathway involving RNase HII. The replicating DNA polymerase (family B or D) of the lagging strand replicates DNA in the presence of PCNA, and RNase HII specifically cuts into the initiating RNA (step I and step II). Removal of the mono-ribonucleotide involves the coordinated activities of DNA synthesis and strand displacement by the replicative DNA polymerase (family B or D), as well as cutting of the displaced strand by FEN1 (step III and step IV). DNA ligase forms the phosphodiester bond between the two adjacent DNA strands (step V); (b) RNase HII-independent reactive pathway. DNA polymerase replicates the lagging strand in the presence of PCNA (step I). Complete removal of the initiator RNA involves the coordinated activities of DNA synthesis and strand displacement by the replicative DNA polymerase (family B or D), and cutting of the displaced strand by FEN1 (step II and step III). DNA ligase forms the phosphodiester bond between the two adjacent DNA strands (step IV).*

The ligation of DNA ends is an essential step during the maturation process of Okazaki fragments. DNA ligase appears to be an essential protein for archaeal survival (Sarmiento et al. 2013). When multiple copies of genes exist in genomes, only their simultaneous genetic deletion is lethal, indicating that functional redundancy is essential for cell survival (Zhao et al. 2006). Biochemical experiments have shown in vitro that DNA ligase plays a significant role in the Okazaki fragment maturation process, beyond its function of ligating DNA strands. Indeed, LIG1 interacts functionally with DNA polymerases and FEN1, stimulating strand-displacement activity and cleavage activity, respectively (Henneke 2012), with ligation taking place after complete removal of primer ribonucleotides. Unlike the Okazaki fragment maturation process in eukaryotes, ligation in archaea can take place close to the RNA–DNA junction, since the DNA fragment downstream of the RNA is synthesized by a high-fidelity DNA polymerase (Henneke et al. 2005).

1.4. Restarting replication forks

Once the replisome components are in place at the replication fork, DNA synthesis and Okazaki fragment maturation take place as described above. However, this process is never as straightforward. In the course of a cell's life, DNA can be subjected to numerous natural stresses, both endogenous and exogenous. In particular, cellular metabolism produces reactive oxygen species, which can cause DNA damages. Similarly, exposure of cells to exogenous genotoxic agents, such as ultraviolet radiation or ionizing radiation, is a source of damage. Numerous repair mechanisms exist, enabling an adapted response to each type of lesion. However, some DNA damages are not repaired prior to replication, which can cause a replication fork to stop at the DNA lesions. Other events, such as collision between the replication fork and the transcription machinery, or the presence of repeated regions of DNA capable of forming secondary structures, can also hinder replication. These various events can lead the replication fork to slowdown or even stop. These events are known as replication stress. Replication stress leads to DNA damage and genetic instability, contributing to the development of cancer in higher eukaryotes. Understanding the molecular basis of the cellular response to replication stress is therefore a major challenge. The decade-long study of the response to replication stress in bacteria, then in unicellular eukaryotes and mammalian cells (Carr and Lambert 2013; Michel and Sandler 2017), has revealed a mechanistic link between replication and recombination. Indeed, the restart of replication forks that have stopped involves proteins of the homologous recombination pathway, which can lead to chromosomal rearrangements at replication arrest sites. These studies have enabled us to understand the molecular basis for the stability of arrested replication forks and their restart, depending on the nature of the replication arrest.

The response to replication stress in archaea is still largely unknown. However, a close link between replication and recombination has also been found in archaea. In hyperthermophilic archaea, key homologous recombination proteins, notably the RadA recombinase responsible for strand invasion, are essential for cell survival (Fujikane et al. 2010). In these organisms, this close link between replication and recombination ensures replication fidelity and stability. Indeed, the mismatch repair pathway present in bacteria and eukaryotes does not appear to be present in hyperthermophilic archaea. It could therefore be that mismatches cause replication fork arrests, whose restart would be ensured by the homologous recombination pathway, thus coupling mismatch repair and fork restart (Grogan 2015) (Figure 1.4, steps c and d). As mentioned at the beginning of this chapter, it has also been suggested, in the hyperthermophilic archaea *T. kodakarensis*, that replication initiation occurs exclusively from homologous recombination intermediates, independently of the presence of an origin of replication associated with a replication initiation protein (Gehring et al. 2017). These intermediates, known as D-loops, correspond to three-branched DNA structures generated after strand invasion by the recombination machinery, onto which the replication machinery can be loaded to restart replication in the event of arrest, or even to initiate replication in the absence of canonical replication initiation at replication origins (Figure 1.4).

Biochemical studies aimed at identifying proteins with activity on branched DNA fork substrates in the hyperthermophilic archaea *P. furiosus* have identified the Hef protein, a helicase/nuclease of the XPF/MUS81/FANCM family. This protein is present in eukaryotes and archaea, but is absent in bacteria. Most eukaryotes possess at least four members, which are involved in DNA repair, meiotic recombination and/or replication restart (Ciccia et al. 2008). The biochemical characterisation of the *P. furiosus hef* protein has proposed a role for Hef in the remodeling of arrested replication forks (Komori et al. 2004) (Figure 1.4, steps b and c). Based on these results, in vivo studies in the mesophilic archaea *H. volcanii* have demonstrated a role for Hef in restarting replication, as well as a role for homologous recombination proteins (Lestini et al. 2010, 2013).

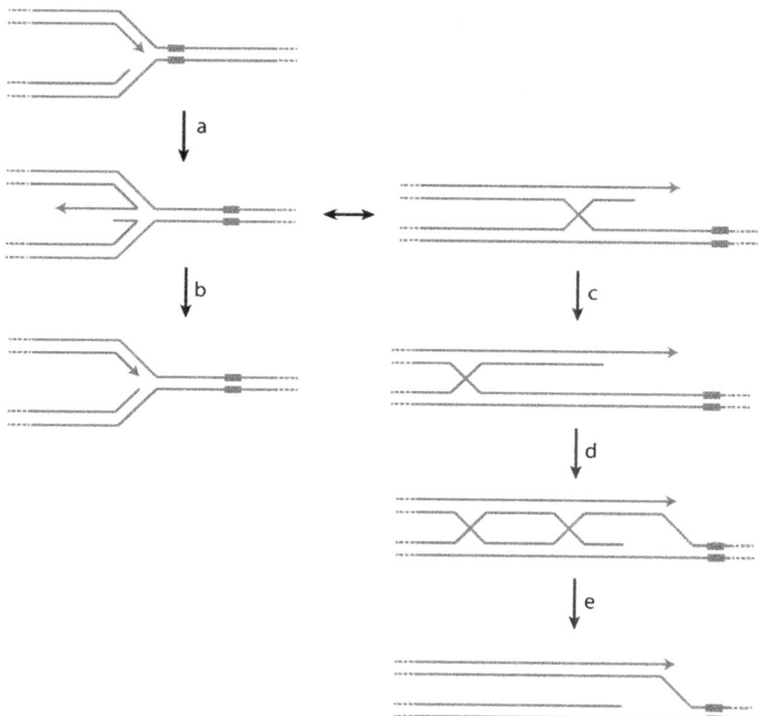

Figure 1.4. *Restarting arrested replication forks . For a color version of this figure, see www.iste.co.uk/clouet/archaea2.zip*

COMMENT ON FIGURE 1.4.– *(a) After replication fork arrest, fork regression leads to pairing of the neosynthesized strands and the formation of a 4-branched DNA structure. A role for the Hef helicase in this remodeling has been proposed (Komori et al. 2004); (b) the action of nucleases (endonuclease or exonuclease), which can be coupled to helicases, restores a three-branch structure on which a replisome can, again, be assembled to enable replication to restart; (c) degradation of the 5′ end by a 5′–3′ exonuclease results in the formation of a single-stranded region with a 3′ end, on which the RadA recombination protein polymerizes to form a filament, while migration of the Holliday junction shifts this four-branch structure; (d) the RadA filament catalyzes the strand invasion reaction, leading to the formation of a second Holliday junction; (e) resolution of the Holliday junctions restores a three-branch structure, known as the D-loop, on which a replisome can again be assembled to enable replication to restart. The neo-synthesized DNA strands are in red, while the template strands are in blue. The arrow represents the 3′ end. Dotted lines represent strand continuity. The blue rectangles on the template strands represent the region where the arrest occurred.*

In archaea, we find a close link between replication and recombination, which seems to be involved in both the initiation of replication and the response to replication stress. A better understanding of these mechanisms in archaea will undoubtedly deepen our understanding of these fundamental processes, which are common to bacteria and eukaryotes alike.

1.5. Replication termination

This stage of replication has hardly been studied in archaea. In *Sulfolobales*, which contain a circular genome with several replication origins, no pause sites have been found between two origins (Duggin et al. 2011). It appears that replication pauses correspond to the meeting of two converging forks, which is a similar case in eukaryotes. Given the non-synchronous origins and the variable size of replicons, only one of these regions corresponds to a genome replication termination.

Due to the circular nature of genomes, during replication, sister chromatids can recombine, giving rise to dimeric chromosomes. It has been shown that there are at least two systems for resolving these dimers, similar to those found in bacteria. In *Sulfolobales*, the presence of a XerD recombinase at the specific *dif* site resolves these dimers into monomers. However, unlike in bacteria, this site is not located close to a termination region, so there is an uncoupling between replication termination and resolution of chromosome dimers (Duggin et al. 2011). In a number of *Euryarcheota*, such as *Thermoplasma acidophilum*, a *dif* site has also been found; however, for these organisms, resolution uses a recombinase from the XerA family (Cortez et al. 2010).

1.6. Chromosome segregation and decatenation

Once again, very little is currently known about the segregation of archaeal chromosomes and their correct distribution in daughter cells. As in all organisms, after replication, the two sister chromatids are joined, requiring decatenation; in other words, the complete separation of the two chromatids. To achieve this, transient double-strand breaks are made by type II topoisomerases. Archaea possess the characteristic subtype IIB, which is capable of decatenation. However, it is possible that type IA topoisomerases, which transiently cut a single strand of DNA, are capable of separating sister chromatids during replication (Couturier et al. 2020).

1.7. References

Ayyagari, R., Gomes, X.V., Gordenin, D.A., Burgers, P.M.J. (2003). Okazaki fragment maturation in yeast. I. Distribution of functions between FEN1 AND DNA2. *J. Biol. Chem.*, 278, 1618–1625.

Balakrishnan, L. and Bambara, R.A. (2013). Flap endonuclease 1. *Annu. Rev. Biochem.*, 82, 119–138.

Berquist, B.R., DasSarma, P., DasSarma, S. (2007). Essential and non-essential DNA replication genes in the model halophilic Archaeon, *Halobacterium* sp. NRC-1. *BMC Genet*, 8(31), 8–31.

Birien, T., Thiel, A., Henneke, G., Flament, D., Moalic, Y., Jebbar, M. (2018). Development of an effective 6-methylpurine counterselection marker for genetic manipulation in *Thermococcus barophilus*. *Genes (Basel)*, 9, E77.

Burkhart, B.W., Cubonova, L., Heider, M.R., Kelman, Z., Reeve, J.N., Santangelo, T.J. (2017). The GAN exonuclease or the flap endonuclease Fen1 and RNase HII are necessary for viability of *Thermococcus kodakarensis*. *J. Bacteriol.*, 199, e00141-17.

Cann, I.K., Komori, K., Toh, H., Kanai, S., Ishino, Y. (1998). A heterodimeric DNA polymerase: Evidence that members of Euryarchaeota possess a distinct DNA polymerase. *Proc. Natl. Acad. Sci. USA*, 95, 14250–14255.

Carr, A.M. and Lambert, S. (2013). Replication stress-induced genome instability: The dark side of replication maintenance by homologous recombination. *J. Mol. Biol.*, 425, 4733–4744.

Chapados, B.R., Hosfield, D.J., Han, S., Qiu, J., Yelent, B., Shen, B., Tainer, J.A. (2004). Structural basis for FEN-1 substrate specificity and PCNA-mediated activation in DNA replication and repair. *Cell*, 116, 39–50.

Ciccia, A., McDonald, N., West S.C. (2008). Structural and functional relationships of the XPF/MUS81 family of proteins. *Annu. Rev. Biochem.*, 77, 259–287.

Clouet-d'Orval, B., Batista, M., Bouvier, M., Quentin, Y., Fichant, G., Marchfelder, A., Maier, L.-K. (2018). Insights into RNA-processing pathways and associated RNA-degrading enzymes in Archaea. *FEMS Microbiol. Rev.*, 42, 579–613.

Cortez, D., Quevillon-Cheruel, S., Gribaldo, S., Desnoues, N., Sezonov, G., Forterre, P., Serre, M.-C. (2010). Evidence for a Xer/dif system for chromosome resolution in archaea. *PLoS Genet*, 6, e1001166.

Couturier, M., Gadelle, D., Forterre, P., Nadal, M., Garnier, F. (2020). The reverse gyrase TopR1 is responsible for the homeostatic control of DNA supercoiling in the hyperthermophilic archaeon *Sulfolobus solfataricus*. *Mol. Microbiol.*, 113, 356–368.

Duggin, I.G., Dubarry, N., Bell, S.D. (2011). Replication termination and chromosome dimer resolution in the archaeon *Sulfolobus solfataricus*. *EMBO J.*, 30, 145–153.

Friedrich-Heineken, E., Henneke, G., Ferrari, E., Hübscher, U. (2003). The acetylatable lysines of human Fen1 are important for endo- and exonuclease activities. *J. Mol. Biol.*, 328, 73–84.

Fujikane, R., Ishino, S., Ishino, Y., Forterre, P. (2010). Genetic analysis of DNA repair in the hyperthermophilic archaeon, *Thermococcus kodakaraensis*. *Genes & Genetic Systems*, 85, 243–257.

Gehring, A.M., Astling, D.P., Matsumi, R., Burkhart, B.W., Kelman, Z., Reeve, J.N., Jones, K.L., Santangelo, T.J. (2017). Genome replication in *Thermococcus kodakarensis* independent of Cdc6 and an origin of replication. *Frontiers in Microbiology*, 8, 2084.

Greci, M.D. and Bell, S.D. (2020). Archaeal DNA replication. *Annu. Rev. Microbiol.*, 74, 65–80.

Grogan, D.W. (2015). Understanding DNA repair in hyperthermophilic archaea: Persistent gaps and other reasons to focus on the fork. *Archaea*, 2015(1), 942605.

Hawkins, M., Malla, S., Blythe, M.J., Nieduszynski, C.A., Allers, T. (2013). Accelerated growth in the absence of DNA replication origins. *Nature*, 503, 544–547.

Heider, M.R., Burkhart, B.W., Santangelo, T.J., Gardner, A.F. (2017). Defining the RNaseH2 enzyme-initiated ribonucleotide excision repair pathway in Archaea. *J. Biol. Chem.*, 292, 8835–8845.

Henneke, G. (2012). In vitro reconstitution of RNA primer removal in Archaea reveals the existence of two pathways. *Biochem. J.*, 447, 271–280.

Henneke, G., Friedrich-Heineken, E., Hübscher, U. (2003). Flap endonuclease 1: A novel tumour suppresser protein. *Trends Biochem. Sci.*, 28, 384–390.

Henneke, G., Flament, D., Hübscher, U., Querellou, J., Raffin, J.-P. (2005). The hyperthermophilic euryarchaeota *Pyrococcus abyssi* likely requires the two DNA polymerases D and B for DNA replication. *J. Mol. Biol.*, 350, 53–64.

Hosfield, D.J., Frank, G., Weng, Y., Tainer, J.A., Shen, B. (1998). Newly discovered archaebacterial flap endonucleases show a structure-specific mechanism for DNA substrate binding and catalysis resembling human flap endonuclease-1. *J. Biol. Chem.*, 273, 27154–27161.

Jacob, F., Brenner, S., Cuzin, F. (1963). On the regulation of DNA replication in bacteria. *Cold Spring Harb. Symp. Quant. Biol.*, 28, 329–348.

Jeong, H.-S., Backlund, P.S., Chen, H.-C., Karavanov, A.A., Crouch, R.J. (2004). RNase H2 of *Saccharomyces cerevisiae* is a complex of three proteins. *Nucleic Acids Res.*, 32, 407–414.

Kelman, L.M. and Kelman, Z. (2014). Archaeal DNA replication. *Annual Review of Genetics*, 48, 71–97.

Komori, K., Hidaka, M., Horiuchi, T., Fujikane, R., Shinagawa, H., Ishino, Y. (2004). Cooperation of the N-terminal helicase and C-terminal endonuclease activities of archaeal Hef protein in processing stalled replication forks. *J. Biol. Chem.*, 279, 53175–53185.

Kushida, T., Narumi, I., Ishino, S., Ishino, Y., Fujiwara, S., Imanaka, T., Higashibata, H. (2019). Pol B, a family B DNA polymerase, in *Thermococcus kodakarensis* is important for DNA repair, but not DNA replication. *Microbes Environ.*, 34, 316–326.

Lai, L., Yokota, H., Hung, L.W., Kim, R., Kim, S.H. (2000). Crystal structure of archaeal RNase HII: A homologue of human major RNase H. *Structure*, 8, 897–904.

Le Breton, M., Henneke, G., Norais, C., Flament, D., Myllykallio, H., Querellou, J., Raffin, J.-P. (2007). The heterodimeric primase from the euryarchaeon *Pyrococcus abyssi*: A multifunctional enzyme for initiation and repair? *J. Mol. Biol.*, 374, 1172–1185.

Lestini, R., Duan, Z., Allers, T. (2010). The archaeal Xpf/Mus81/FANCM homolog Hef and the Holliday junction resolvase Hjc define alternative pathways that are essential for cell viability in *Haloferax volcanii. DNA Repair (Amst)*, 9, 994–1002.

Lestini, R., Laptenok, S.P., Kühn, J., Hink, M.A., Schanne-Klein, M.-C., Liebl, U., Myllykallio, H. (2013). Intracellular dynamics of archaeal FANCM homologue Hef in response to halted DNA replication. *Nucleic Acids Res.*, 41, 10358–10370.

Liu, B., Ouyang, S., Makarova, K.S., Xia, Q., Zhu, Y., Li, Z., Guo, L., Koonin, E.V., Liu, Z.-J., Huang, L. (2015). A primase subunit essential for efficient primer synthesis by an archaeal eukaryotic-type primase. *Nat. Commun.*, 6, 7300.

Lundgren, M. and Bernander, R. (2007). Genome-wide transcription map of an archaeal cell cycle. *Proc. Natl. Acad. Sci. USA*, 104, 2939–2944.

Lundquist, R.C. and Olivera, B.M. (1982). Transient generation of displaced single-stranded DNA during nick translation. *Cell*, 31, 53–60.

Malfatti, M.C., Henneke, G., Balachander, S., Koh, K.D., Newnam, G., Uehara, R., Crouch, R.J., Storici, F., Tell, G. (2019). Unlike the *Escherichia coli* counterpart, archaeal RNase HII cannot process ribose monophosphate abasic sites and oxidized ribonucleotides embedded in DNA. *J. Biol. Chem.*, 294, 13061–13072.

Matsui, E., Musti, K.V., Abe, J., Yamasaki, K., Matsui, I., Harata, K. (2002). Molecular structure and novel DNA binding sites located in loops of flap endonuclease-1 from *Pyrococcus horikoshii. J. Biol Chem.*, 277, 37840–37847.

Matsunaga, F., Forterre, P., Ishino, Y., Myllykallio, H. (2001). In vivo interactions of archaeal Cdc6/Orc1 and minichromosome maintenance proteins with the replication origin. *Proc. Natl. Acad. Sci. USA*, 98, 11152–11157.

Matsunaga, F., Takemura, K., Akita, M., Adachi, A., Yamagami, T., Ishino, Y. (2009). Localized melting of duplex DNA by Cdc6/Orc1 at the DNA replication origin in the hyperthermophilic archaeon *Pyrococcus furiosus. Extremophiles*, 14, 21–31.

Meslet-Cladiére, L., Norais, C., Kuhn, J., Briffotaux, J., Sloostra, J.W., Ferrari, E., Hübscher, U., Flament, D., Myllykallio, H. (2007). A novel proteomic approach identifies new interaction partners for proliferating cell nuclear antigen. *J. Mol. Biol.*, 372, 1137–1148.

Michel, B. and Sandler, S.J. (2017). Replication restart in bacteria. *J. Bacteriol.*, 199, e00102-17.

Neelsen, K.J. and Lopes, M. (2015). Replication fork reversal in eukaryotes: From dead end to dynamic response. *Nat. Rev. Mol. Cell. Biol.*, 16, 207–220.

Norais, C., Hawkins, M., Hartman, A.L., Eisen, J.A., Myllykallio, H., Allers, T. (2007). Genetic and physical mapping of DNA replication origins in *Haloferax volcanii*. *PLoS Genet*, 3, e77.

Ogawa, T., Pickett, G.G., Kogoma, T., Kornberg, A. (1984). RNase H confers specificity in the dnaA-dependent initiation of replication at the unique origin of the *Escherichia coli* chromosome in vivo and in vitro. *Proc. Natl. Acad. Sci. USA*, 81, 1040–1044.

Okazaki, R., Okazaki, T., Sakabe, K., Sugimoto, K., Sugino, A. (1968). Mechanism of DNA chain growth. I. Possible discontinuity and unusual secondary structure of newly synthesized chains. *Proc. Natl. Acad. Sci. USA*, 59, 598–605.

Pascal, J.M., O'Brien, P.J., Tomkinson, A.E., Ellenberger, T. (2004). Human DNA ligase I completely encircles and partially unwinds nicked DNA. *Nature*, 432, 473–478.

Pascal, J.M., Tsodikov, O.V., Hura, G.L., Song, W., Cotner, E.A., Classen, S., Tomkinson, A.E., Tainer, J.A., Ellenberger, T. (2006). A flexible interface between DNA ligase and PCNA supports conformational switching and efficient ligation of DNA. *Mol. Cell*, 24, 279–291.

Robinson, N.P., Blood, K.A., McCallum, S.A., Edwards, P.A.W., Bell, S.D. (2007). Sister chromatid junctions in the hyperthermophilic archaeon *Sulfolobus solfataricus*. *EMBO J.*, 26, 816–824.

Rolland, J., Gueguen, Y., Persillon, C., Masson, J.-M., Dietrich, J. (2004). Characterization of a thermophilic DNA ligase from the archaeon *Thermococcus fumicolans*. *FEMS Microbiol. Lett.*, 236, 267–273.

Sarmiento, F., Mrázek, J., Whitman, W.B. (2013). Genome-scale analysis of gene function in the hydrogenotrophic methanogenic archaeon *Methanococcus maripaludis*. *Proc. Natl. Acad. Sci. USA*, 110, 4726–4731.

Sato, A., Kanai, A., Itaya, M., Tomita, M. (2003). Cooperative regulation for Okazaki fragment processing by RNase HII and FEN-1 purified from a hyperthermophilic archaeon, *Pyrococcus furiosus*. *Biochem. Biophys. Res. Commun.*, 309, 247–252.

Sauguet, L., Raia, P., Henneke, G., Delarue, M. (2016). Shared active site architecture between archaeal PolD and multi-subunit RNA polymerases revealed by X-ray crystallography. *Nat. Commun.*, 7, 12227.

Sriskanda, V. and Shuman, S. (1998). Mutational analysis of chlorella virus DNA ligase: Catalytic roles of domain I and motif VI. *Nucleic Acids Res.*, 26, 4618–4625.

Stodola, J.L. and Burgers, P.M. (2016). Resolving individual steps of Okazaki fragment maturation at msec time-scale. *Nat. Struct. Mol. Biol.*, 23, 402–408.

Storici, F., Henneke, G., Ferrari, E., Gordenin, D.A., Hübscher, U., Resnick, M.A. (2002). The flexible loop of human FEN1 endonuclease is required for flap cleavage during DNA replication and repair. *EMBO J.*, 21, 5930–5942.

Tadokoro, T. and Kanaya, S. (2009). Ribonuclease H: Molecular diversities, substrate binding domains, and catalytic mechanism of the prokaryotic enzymes. *FEBS J.*, 276, 1482–1493.

Taib, N., Gribaldo, S., MacNeill, S.A. (2021). Single-stranded DNA-binding proteins in the archaea. *Methods Mol. Biol.*, 2281, 23–47.

Vijayakumar, S., Chapados, B.R., Schmidt, K.H., Kolodner, R.D., Tainer, J.A., Tomkinson, A.E. (2007). The C-terminal domain of yeast PCNA is required for physical and functional interactions with Cdc9 DNA ligase. *Nucleic Acids Res.*, 35, 1624–1637.

Yan, J., Holzer, S., Pellegrini, L., Bell, S.D. (2018). An archaeal primase functions as a nanoscale caliper to define primer length. *Proceedings of the National Academy of Sciences*, 115, 6697–6702.

Yang, H., Wu, Z., Liu, J., Liu, X., Wang, L., Cai, S., Xiang, H. (2015). Activation of a dormant replication origin is essential for *Haloferax mediterranei* lacking the primary origins. *Nat. Commun.*, 6, 8321.

Zhao, A., Gray, F.C., MacNeill, S.A. (2006). ATP- and NAD+-dependent DNA ligases share an essential function in the halophilic archaeon *Haloferax volcanii*. *Mol. Microbiol.*, 59, 743–752.

2

Archaeal DNA Repair

Caroline L'HERMITTE-STEAD, Anaïs BAYARD, Alexey ALEKSANDROV, Roxane LESTINI and Hannu MYLLYKALLIO

Laboratoire d'optique et biosciences, Inserm, CNRS, Institut Polytechnique de Paris, Palaiseau, France

2.1. General introduction

All cellular organisms, including archaea, and many viruses have adapted DNA as the medium to stock genetic information. This choice is a double-edged sword. However, accurate replication of the genetic information is required for maintaining genetic stability between generations, stable changes in DNA sequence (mutations) also drive evolution and underpin microbial adaptation to rapidly changing environmental conditions. Relatively frequent changes in the sequence or the chemical structure of DNA potentially modify the information stored in the sequence of DNA duplex (Lindahl 1993). However, dedicated DNA repair machines efficiently repair the vast majority of this potential misinformation, thus limiting the passage of erroneous genetic information to the next generations.

The specific goal of this chapter on *archaeal DNA repair* is to outline how nature has invented several different strategies, DNA repair pathways, to detect and correct various types of DNA damages. The DNA damage can impact either the single strand or both strands of the DNA duplex.

Biology of Archaea 2,
coordinated by Béatrice CLOUET-D'ORVAL, Bruno FRANZETTI and Philippe OGER.

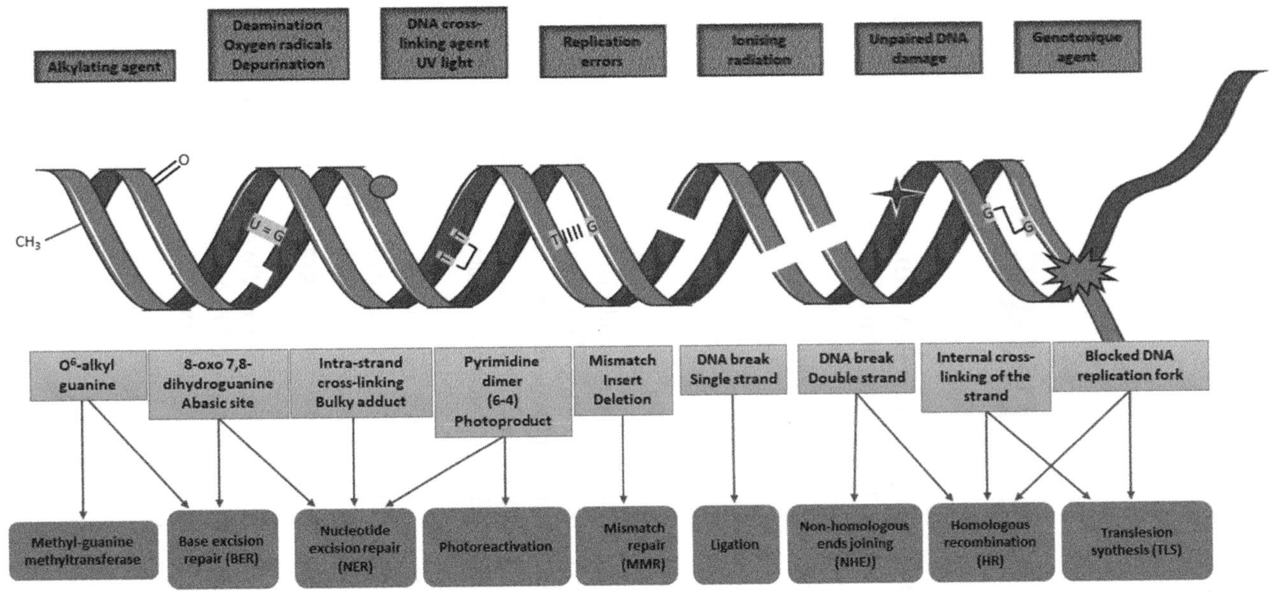

Figure 2.1. *The different types of abnormal sites including replication errors (mismatches) and sites of chemical DNA damage are indicated within the double-stranded DNA molecule. These sites are either efficiently repaired by highly specific molecular machines indicated in the figure, or, if left unrepaired, may result in the formation of mutations, which are passed between the generations. This figure was redrawn after White and Allers (2018). For a color version of this figure, see www.iste.co.uk/clouet/archaea2.zip*

In the direct DNA repair process, chemical modifications of DNA are removed without cutting the DNA backbone. This is the case for the repair of photoproducts caused by UV radiation that are directly repaired using the enzyme photolyase. Erroneous genetic information is frequently repaired by the different excision repair pathways in archaea, which include mismatch repair (MMR), base excision repair (BER), nucleotide excision repair (NER) and ribonucleotide excision repair (RER) pathways. Double-strand breaks (DSB) may be precisely repaired by homologous recombination (HR), which uses the intact DNA strand as a backup for the correct genetic information (for detailed reviews on archaeal DNA repair, see, for example, White and Allers 2018).

In humans, the crucial importance of DNA repair for cellular physiology is exemplified by the fact that they are often linked to disease and/or required for optimal progression of the cell cycle. In the case of archaea, the importance of the accurate maintenance of genetic information is evident from the fact that the substantial portion of the predicted and widely distributed archaeal ancestral clusters of orthologous genes (COGs) encode for the proteins required for DNA replication, DNA repair, transcription and translation (Figure 2.2).

Notably, a sizable portion of the genome for the archaeal ancestor may encode for DNA replication and/or repair functions, thus indicating that the sophisticated DNA replication systems and repair pathways ensuring high fidelity DNA replication were present already in the last archaeal common ancestor (LACA). In comparison to DNA replication, higher error rates are tolerated for RNA and protein synthesis as the information carried by the latter group of the biological polymers is not passed between the generations.

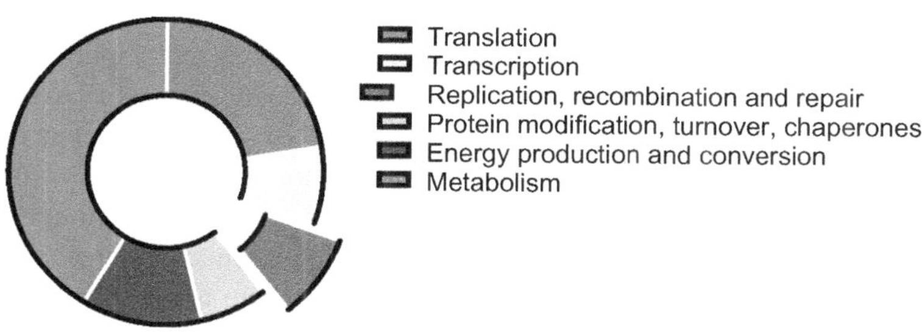

Figure 2.2. *Early evolutionary genomics analysis of archaeal genomes indicated that up to 10% of ancestral archaeal genes may participate in genome replication and repair in the LACA. This figure was based on the analysis of 41 archaeal genomes including a total of 681 gene clusters, as indicated by Makarova et al. (2017). For a color version of this figure, see www.iste.co.uk/clouet/archaea2.zip*

2.2. Several different types of DNA damages occur in archaea

Information content of archaeal genomes is constantly challenged and attacked by a wide range of intrinsic and external causes, which will be summarized in this section.

2.2.1. *The archaeal replisome as the intrinsic source of replication errors and DNA damage*

The archaeal replisome is the complex molecular machinery that faithfully copies genetic information carried by DNA, thus allowing the inheritance of biological information between generations. The high accuracy of the replisome, and DNA polymerase, in particular, is facilitated by several distinct mechanisms. Incorporation of the correct nucleotide during DNA replication is guaranteed by the selection of the correct base and proof-reading activities of DNA polymerases during DNA polymerization. In archaea, most work has focused on the biochemical characterization of the DNA polymerase families B and D (reviewed in Cooper 2018), the latter discovered by the Ishino laboratory and only found in the archaea (Cann et al. 1998). These polymerases differentiate between correct and incorrect base pairs by their differential binding to the active site. This initial and highly discriminating binding event is followed by a catalytic step that guarantees the formation of the phosphodiester bond and therefore the continuity of the DNA strand. This nucleotide selectivity is a sum of several factors including the hydrogen bonding between template bases and incoming dNTPs by excluding interactions with bound water molecules. The size and the shape complementarity between the correct nascent base pairs and the active site pocket also plays an important role. Both Watson–Crick base pairs A-T and G-C are formed by pairing purine and pyrimidine bases. Consequently, these canonical base pairs are highly similar in terms of their size and shape, which greatly facilitates their selective recognition over mispaired bases. Occasionally, DNA polymerase nevertheless makes mistakes in a base selection, resulting in erroneous incorporation of wrong nucleotides to a growing DNA chain. To deal with these potential DNA replication mistakes, archaeal replicative DNA polymerases use 3′ to 5′ exonuclease (proofreading) activity to remove the mispaired nucleotide before the chain elongation takes place, further increasing the accuracy of DNA synthesis.

Despite these sophisticated built-in fidelity mechanisms, archaeal DNA polymerases occasionally create replication errors and DNA damage within the cell. A typical error rate for DNA polymerases is $\approx 10^{-4}$–10^{-5}. The overall fidelity of DNA synthesis can be improved by the proofreading activity by an approximate factor of 10^{2}–10^{3}, suggesting that the wrong deoxynucleotide is approximately incorporated once every 10^6–10^8 nucleotide. DNA polymerases also incorporate surprisingly frequently *ribonucleotides* (rNTPs) into chromosomal DNA, with an error rate of ≈ 10^{-3}. This is somewhat surprising as archaeal DNA polymerases carry specific structural features that block the binding of rNTPs into their active site. However, the intracellular concentrations of rNTPs are markedly higher than dNTPs in the archaeal cell, which contributes to a high frequency of ribonucleotides that are embedded within the duplex DNA up to every 1 kb. This decreases the chemical stability of archaeal genomes, thus challenging the accuracy of genetic information. archaeal cells must use specific strategies to remove misincorporated bases and sugar moieties from DNA before passing genetic information to the next generation.

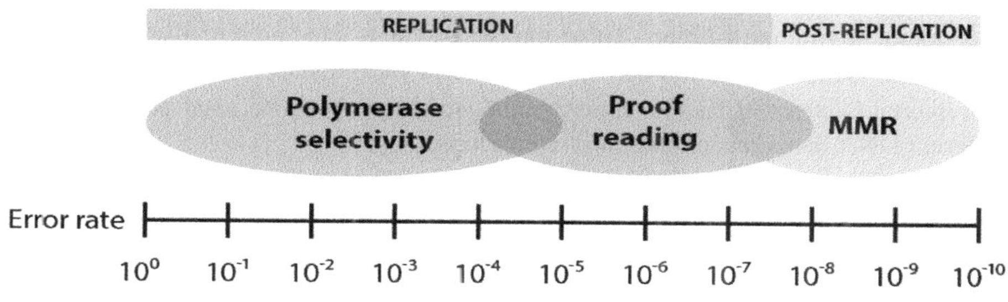

Figure 2.3. *Different mechanisms ensuring high fidelity of DNA replication in archaea (presented using a logarithmic scale). Base selection and proofreading are performed during the DNA replication by DNA polymerases, whereas the MMR takes place in a post-replicative manner by specific protein complexes. For a color version of this figure, see www.iste.co.uk/clouet/archaea2.zip*

2.2.2. Chemical and physical factors threaten the integrity of archaeal genomes

DNA molecule is a complex organic chemical that is, in addition to replication errors, susceptible to change through intracellular chemical reactions resulting in decomposing of genetic information. Major chemical challenges that attack the chemical integrity of DNA within cells are hydrolytic attacks and oxidative DNA damage (Figure 2.1). DNA is also frequently chemically modified by activated intracellular carbon donors like S-adenosylmethionine, resulting in alkylation (methylation) of DNA (Lindahl 1993).

Hydrolytic reactions refer to chemical reactions where water molecules act as a catalyst to break chemical bonds, thus resulting in the splitting of the large organic molecules like DNA to their smaller constituents. The importance of hydrolytic reactions, attacking DNA, is amplified by a very high intracellular concentration of water, which can approach 55 M. This omnipresent water favors hydrolytic attacks of the chemical bonds linking DNA bases to sugars (Figure 2.1), thus resulting in either the removal of purines (depurination) or pyrimidine (depyrimidination) bases from the DNA duplex. Hydrolytic reactions can also readily remove an amino group from the cytosine base, particularly from ssDNA, thus resulting in cytosine deamination. Reactive oxygen species (*oxidative DNA damage*) also has a high potential to chemically modify the structure of DNA, resulting for instance in DNA fragmentation and formation of 8-Oxoguanine, one of the most common DNA lesions. In addition to these chemical factors, ultraviolet and gamma radiation can drastically modify the structure of the DNA. UV can result in the formation of interstrand thymidine dimers (photoproducts) that can potentially halt DNA replication. Gamma radiation can also create ssDNA or dsDNA breaks that need to be repaired. Several different experimental conditions can be used to mimic the different DNA damages in laboratory conditions.

Archaea frequently thrive in extreme conditions such as high temperature, salt, pressure or are exposed to ultraviolet or gamma radiation. All of these environmental conditions potentially accelerate the formation of the different DNA lesions indicated in Figure 2.1. For instance, high temperatures can accelerate DNA degradation and cytosine deamination by the factor of 3,000 between 37°C and 100°C. Nevertheless, archaeal mutation rates are similar to mesophilic bacteria, indicating that very efficient DNA repair pathways must exist in archaea.

Compound/conditions	Main DNA damages induced	Implicated DNA repair pathway
UV	Cyclobutane-pyrimidine dimers (CPDs) and pyrimidine adducts (6-4PPs)	Photolyase (direct repair) or nucleotide excision repair (NER)
Gamma radiation Phleomycin	DNA double-strand breaks (DSBs)	Homologous recombination
Alkylating agents (hydrogen peroxide, methyl methanesulfonate)	Oxidized and methylated bases	Base-excision repair (BER)
Mitomycin C	Intra- and inter-strand crosslinks (ICL)	ICL-repair

Table 2.1. *Experimental conditions used in the laboratory to mimic different DNA damages occurring in nature*

2.3. Different archaeal DNA repair pathways

In this section, the key DNA repair pathways operating in archaea will be described.

2.3.1. *DNA mismatches threaten genome integrity in all domains of life*

Replication errors often result in the formation of mismatches (MM) that are formed via the base pairing of specific chemical forms of bases (tautomers). As the G: T MM mimics very well naturally occurring Watson–Crick base pairs, DNA polymerase does not readily eliminate them as errors during base selection or proofreading (Kunkel and Erie 2015). However, these mispairings are efficiently removed by the MMR, as will be described below. Nonfunctional MMR will lead to a 100- to 1000-fold increase in mutation rates, exemplifying the importance of the post-replicative MMR. The action of mutagenic chemicals, ionizing radiation or spontaneous deamination may also create mismatches, independently of DNA replication. Genetic studies have also indicated that MMR, at least in some bacteria, also inhibits recombination of closely related but nonidentical sequences, thus acting as an anti-recombinase. This implies that mispairings between two DNA bases may also occur during heteroduplex formation during recombination that are rejected by the MMR. In some cases, MMR also participates in the repair of insertions/deletions, in addition to base mismatches.

Nature has invented two independent solutions for detecting and correcting these mismatches (Figure 2.4) (Kunkel and Erie 2015). The canonical MMR system consists of MutS and MutL proteins that, respectively, recognize and excise the MMs during the archaeal MMR process (Minobe et al. 2019). The binding of the nucleotide-free (or ADP-bound) MutS protein to mispairs initiates the MMR. In the presence of ATP, MutS is thought to bend, thus promoting the formation of the MutSL nuclease complex required for excising the mismatched nucleotide(s). The complex DNA repair process is subsequently completed by a template-dependent resynthesis of the eliminated DNA strand. Canonical post-replicative MMR discriminates between wrong (newly synthesized DNA) and correct (template DNA) genetic information. This can be achieved by asymmetric loading of the MutL endonuclease, which is facilitated by the replication clamp PCNA. The strand specificity signal likely corresponds to DNA strand discontinuities ("nicks") created by Okazaki fragment processing and/or elimination of misincorporated ribonucleotides. However, *Escherichia coli* and related gamma-proteobacteria use a more complex methyl-directed MMR system, facilitated by MutH protein to achieve strand-specific DNA repair (not shown in Figure 2.4).

Despite the crucial functional role of the MMR, early bioinformatics studies surprisingly failed to identify the canonical *mutSL* system in many archaea and actinobacteria. However, a combination of genetic and biochemical studies led to the discovery of a noncanonical mismatch-specific endonuclease NucS (also dubbed EndoMS) that is found in many archaea and actinobacteria (Nakae et al. 2016). These proteins do not show sequence or structural similarity to previously characterized MMR proteins. Genetic inactivation of *nucS* raises the spontaneous rate of mutations by ≈ 200-fold (Castañeda-Garcia et al. 2017; Ishino et al. 2018). This increase in mutation rate is similar to what has been observed in MMR-deficient *E. coli*, indicating that the in vivo efficiencies of both MMR systems are comparable. Recent studies have established that NucS endonucleases function very differently in comparison to MutSL complexes since (i) the NucS-replication clamp complex alone recognizes and cleaves MMs (preferentially G/T, T/T, G/G in this order); (ii) ATP activation of NucS is unnecessary for DNA repair; and (iii) NucS proteins remove MM by creating a highly recombinogenic dsDNA break (DSB) with a cohesive end centered at the mismatch.

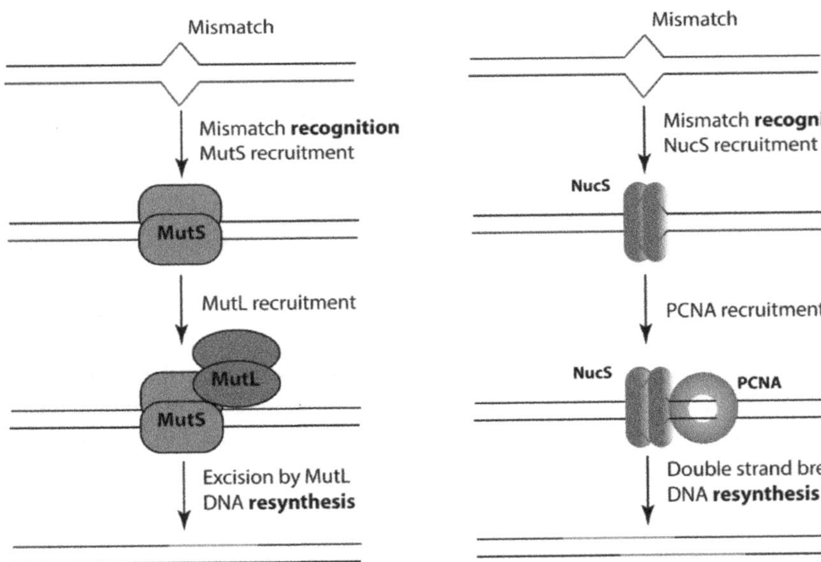

Figure 2.4. *Two different strategies for repairing mismatches that occur during DNA replication. The MM is recognized either by MutS or NucS/EndoMS proteins. While MutS recruits an additional endonuclease MutL to the MM, NucS/EndoMS together with the replication clamp PCNA can also remove the MM using its nuclease activity. For a color version of this figure, see www.iste.co.uk/clouet/archaea2.zip*

Moreover, the formation of the NucS-replication clamp complex is necessary for mutation avoidance suggesting that the tight physical interaction between NucS and the replication clamp is mediated by an evolutionarily conserved peptide motif ("clamp interaction peptide") located at the C-terminus of the NucS orthologs (Ishino et al. 2018). The NucS system does not likely require identification of template (nascent) DNA strand, as the repair is likely performed by incoming dsDNA via HR (or by error-prone NHEJ or microhomology-dependent repair). Genetic studies have also indicated that similar to the MutSL system, NucS also acts as anti-recombinase.

Archaeal species that have MutS and MutL homologs are generally halophiles or methanogens who have likely recovered these genes by gene transfer from bacteria. Recently, it has been proposed that the eukaryotic nuclear MutSL homologs originated within the Asgard archaea, while those functioning in mitochondria or chloroplasts were transferred from alpha-proteobacteria or Cyanobacteria to Eukarya, respectively. On the other hand, NucS/EndoMS homologs are frequently found in thermophilic archaea.

2.3.2. *Archaeal NER*

The NER pathway repairs many different types of DNA damages that distort the helix by locally destabilizing the DNA duplex. Examples of these types of damages include the "dark repair" of DNA photoproducts (when not directly repaired by photolyases) and base modifications created by oxidative stress or endogeneous or exogenous chemical agents. These distinct DNA modifications have a great potential to halt essential DNA processes in the cell, such as DNA replication and transcription. Consequently, the efficient repair of helix distorting DNA damages is crucial for the accurate maintenance of archaeal genetic information (Rouillon and White 2011).

Two different variants of general NER mechanisms have been described. In a global genomic NER (GG-NER), various enzymes recognize DNA distortions, which results in the strand-specific excision of the DNA damage (Figure 2.5). The removal of the damaged genetic information is followed by the resynthesis of DNA using the undamaged strand as the template. The transcription-coupled NER (TCR-NER) removes those DNA lesions that arrest RNA polymerase. Consequently, TRC-NER is specific for DNA template strand that is actively being transcribed by RNA polymerase. In the case of TCR-NER, the stalling of RNA polymerase provides a signal that initiates NER by recruitment of NER proteins, facilitated by transcription repair coupling factors, to the site of DNA damage. No homologs for eukaryotic or bacterial transcription coupling factors CSB or Mfd in archaeal. However, the protein Eta with similar characteristics to Mfd has been recently described in *Thermococcales*, which may function in TCR-NER in some archaea (Walker et al. 2017).

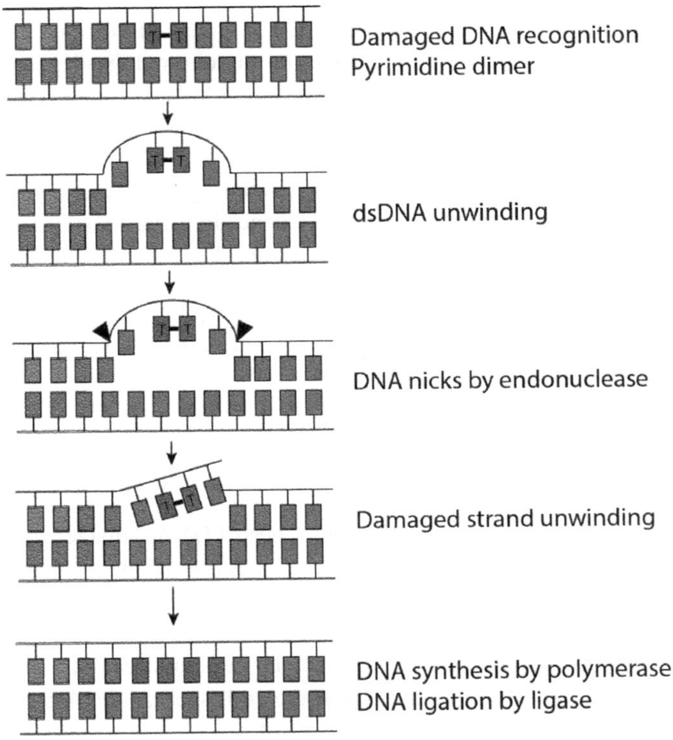

Figure 2.5. *An overview of a global genomic NER (GG-NER) where various enzymes can recognize DNA distortions such as photoproducts (pyrimidine dimers). This initiates the unwinding of the DNA duplex around the damaged site and facilitates the strand-specific excision of the DNA damage to remove erroneous/missing genetic information. DNA polymerases and ligases are then used to reestablish the continuity of the DNA damaged strand. For a color version of this figure, see www.iste.co.uk/clouet/archaea2.zip*

Perhaps the best-characterized NER system in archaea is that of *Haloferax volcanii* (for enzymatic steps, see Figure 2.5). This and other halophiles essentially use bacterial UvrABC proteins that are found in a minority of archaeal species (Perez-Arnaiz et al. 2020). These proteins recognize the bulky DNA damages (UvrA) and locally unwind DNA duplex around this site DNA (UvrB). The DNA melting around the DNA damage then facilitates excision of the bulky adduct by UvrC and UvrD, followed by DNA synthesis and ligation to ensure the continuity of the excised DNA strand. The deletion of these genes in halophiles results in marked UV sensitivity. This observation indicates the central role of these genes in the haloarchaeal NER. As these species are frequently exposed to UV radiation in their natural environment, this is likely physiologically highly significant. Phylogenetic analysis of NER proteins revealed that homologs of the bacterial UvrABC proteins are found only in some mesophilic archaea, which probably acquired them by horizontal gene transfer from bacteria.

In comparison to bacteria and halophiles, the eukaryotic NER uses much more complex machinery to repair the same lesions. In most archaea, some eukaryotic NER-like nucleases (XPF and Fen-1/XPG) and helicases (XPB and XPD) homologs, together with an archaea-specific nuclease, Bax, have been identified, suggesting the existence of at least a partial eukaryotic-like NER pathway in archaea. However, the complete eukaryotic-like XP system has not yet been described in any archaeal species. In addition, despite thorough biochemical, structural, and genetic studies on archaeal XP-like proteins, their direct role in archaeal NER is not evident. Notably, damage recognition or damage verification proteins, including homologs of XPA and XPC, appear to be absent in archaea and, perhaps surprisingly, XPF has been implicated in the restart of archaeal replication forks. Consequently, the identity of the archaeal NER pathway analogous to the eukaryotic one, if it exists, has not yet fully been established. The possibility remains that the possible lack of the complete NER pathway has been compensated by the other functional DNA repair pathways in the majority of archaeal species.

2.3.3. *BER in archaea*

BER is primarily involved in the repair of damaged DNA bases that do not strongly affect the structure of the DNA helix. Examples of these include oxidized and alkylated bases that are subjected to BER. Consequently, BER-deficient archaeal strains are expected to be sensitive to oxidative stress or alkylating agents. Moreover, uracil bases found within DNA, resulting for instance from cytosine deamination, are repaired using this pathway. Note that extreme conditions where archaea frequently thrive can considerably speed up chemical modifications of the DNA bases.

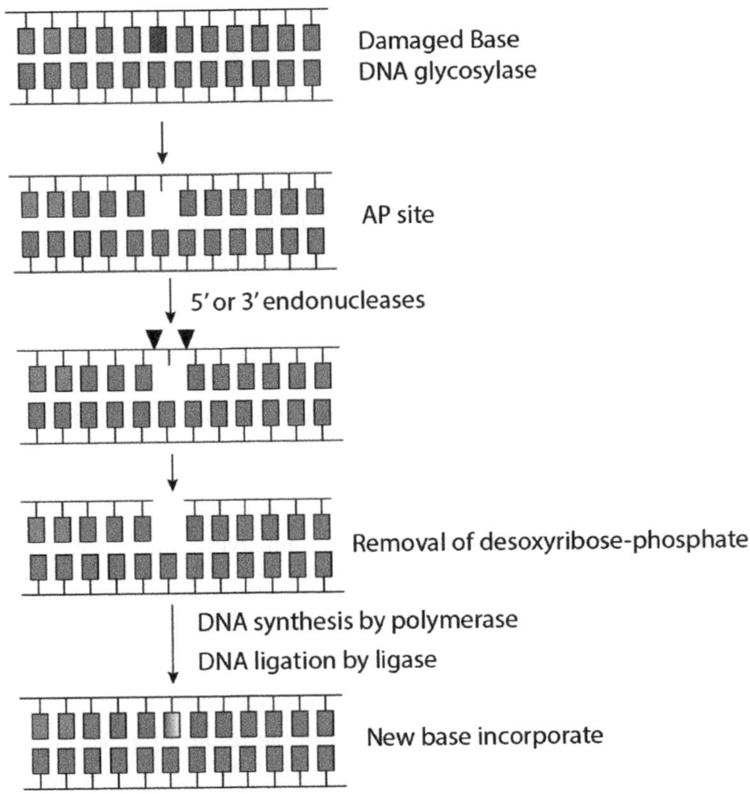

Figure 2.6. *During the short-patch BER, a monofunctional DNA glycosylase cleaves the glycosidic bond between the base and phosphodiester backbone, which results in the formation of an apurinic/apyrimidinic (AP) site. This AP site is further processed by AP-specific lyases/endonucleases, resulting in the formation of a DNA nick. DNA polymerase and ligase activities are used to establish correct genetic information. For a color version of this figure, see www.iste.co.uk/clouet/archaea2.zip*

Canonical archaeal BER (Figure 2.6) (Grasso and Tell 2014) is initiated by glycosylases that specifically recognize oxidized (OGG), alkylated (AlkA) or deaminated bases like uracil (Udg). Two types of glycosylases have been described. Monofunctional enzymes cleave the glycosidic bond between the base and phosphodiester backbone, which results in the formation of an apurinic/apyrimidinic (AP) site. This AP site is then further processed by AP-specific lyases/endonucleases, which allows for the further processing of the damaged bases. A multistep process then allows us to establish the correct genetic information using DNA polymerase and ligase activities. One of the best characterized archaeal glycosylases is the Ogg-family of proteins that act as bifunctional BER enzymes (Gehring et al. 2020). This means that they possess both glycosylase and AP lyase activities for the removal of the base and the cleavage of the phosphodiester backbone. The AP lyase activity creates the single-strand break that can be processed either by a single nucleotide (short-patch BER) or by resynthesis of 2–10 new nucleotides (long-patch BER). The 5′ flap creating during the long-patch BER is cleaved by the Fen1 endonuclease.

In some archaea, alternative endonucleases, very different from the canonical glycosylases, can initiate BER. Endonuclease Q (EndoQ) and endonuclease V (EndoV) cut 5′ and 3′ of damaged bases (Ishino et al. 2015). Of these enzymes, EndoQ shows specificity for uracil (resulting from cytosine deamination), hypoxanthine (oxidative deamination of adenine) and abasic sites. Within archaea, EndoQ is mainly found in *euryarcheota*, particularly in hyperthermophiles and some methanogens. Many enzymes of the BER pathway are well conserved in all three domains of life, indicating the fundamental and general importance of this DNA repair pathway.

2.3.4. *Archaeal RER*

As stated in the introduction, archaeal DNA polymerases misincorporate ribonucleotides, instead of deoxyribonucleotides, into genomic DNA surprisingly frequently (≈ 1 rNTP every 1,000 bases). If these ribonucleotides would remain in DNA, they could potentially distort the structure and stability of the DNA helix. To prevent the potential increase of mutation rate and reduced genome stability, these misincorporated ribonucleotides are removed by the RER pathway. RER is initiated by Rnase*H2* that nicks DNA 5′ of the rNMP embedded in the genomic DNA. The additional enzymes that participate in the archaeal RER include DNA polymerase (displacement of 3′ extremity), Fen1 (cleavage of the flap formed during the process) and DNA ligase, which seals the remaining gap in the DNA (Heider et al. 2017). The role of these enzymes in RER and processing of Okazaki fragments is supported by biochemical reconstruction studies using archaeal proteins.

2.3.5. *Double-strand break repair in archaea*

DSBs are particularly dangerous DNA lesions and a major threat to genomic stability. By simultaneously cutting both DNA strands, they block DNA replication and transcription. Failure to repair DSBs can lead to chromosomal loss, chromosomal rearrangement, cell death or abnormal proliferation. Thus, organisms have evolved several pathways for the repair of DSBs. Archaeal DSBs repair involves three main pathways: the HR repair, the nonhomologous end joining (NHEJ) and microhomology-mediated end joining (MMEJ) (reviewed in White and Allers 2018). HR repair includes the invasion of an undamaged DNA molecule by a damaged molecule of an identical sequence. Resynthesis of the damaged region is accomplished using the undamaged molecule as a template. HR repair can take place in cells with at least two chromosomal copies (e.g. after replication in monoploid or polyploid species). Because an undamaged template strand is available as a template for repair, HR-mediated repair is considered error free. Two alternative pathways relying on the rejoining of the two broken DNA ends exist and have been mostly studied in eukaryotes. The NHEJ directly joins both ends of the DSB and is relatively accurate, although insertions or deletions can occur. MMEJ use microhomologous sequences to align the broken strands. MMEJ is associated with deletions and insertions at the original break site, as well as chromosome translocations.

HR can be divided into three stages: presynapsis, synapsis and post-synapsis. During presynapsis, DSB ends are recognized and processed to produce single-stranded 3′ overhangs. The current archaeal model suggests that DSBs are recognized by the Mre11/Rad50 complex that initiates resection through its 3′–5′ exonuclease activity. This is followed by the recruitment of HerA-NurA complex responsible for the efficient formation of the 3′ overhangs by coupling the translocation activity of HerA, a bipolar helicase, to the degradation of single-strand DNA with a 5′ to 3′ polarity by the NurA nuclease. In hyperthermophilic archaea, the Mre11/Rad50 genes are commonly encoded in operons with both HerA and NurA, implying an early functional link in archaeal DNA ends resection. During synapsis, the 3′ ssDNA extremities are recognized by the conserved recombinase RadA (bacterial RecA and eukaryotic rad51) that polymerizes along the ssDNA region. The dynamic RadA nucleoprotein filament then searches for a homologous sequence, invading dsDNA molecules. The binding of the RadA filament to its homologous sequence in the invaded molecule results in the formation of an intermediate structure referred to as a "D-loop". For the post-synapsis stage, there are two pathways. The invading strand in the D-loop, after priming DNA synthesis, can be ejected from the invaded DNA and annealed with the second end. This process is called synthesis-dependent strand annealing (SDSA). The unwinding of the newly synthesized strand is facilitated by the helicase Hel308. It leads to a localized conversion without crossover. Alternatively, the capture of the second end leads to the formation of a four-branched DNA structure referred to as a Holliday junction (HJ) that must be resolved to complete repair. Archaeal HJ resolvase Hjc recognizes four-way junctions of DNA and uses nuclease activity to resolve the junction. HJs resolution can produce either non-crossover dsDNA products or crossover dsDNA products resulting in a swap of genetic material between the broken DNA molecule and the invaded one. Post-translational modifications of DSB repair components, including methylation of Mre11/Rad50 or phosphorylation of Hjc, likely contribute to the efficiency and rates of the different HR pathways.

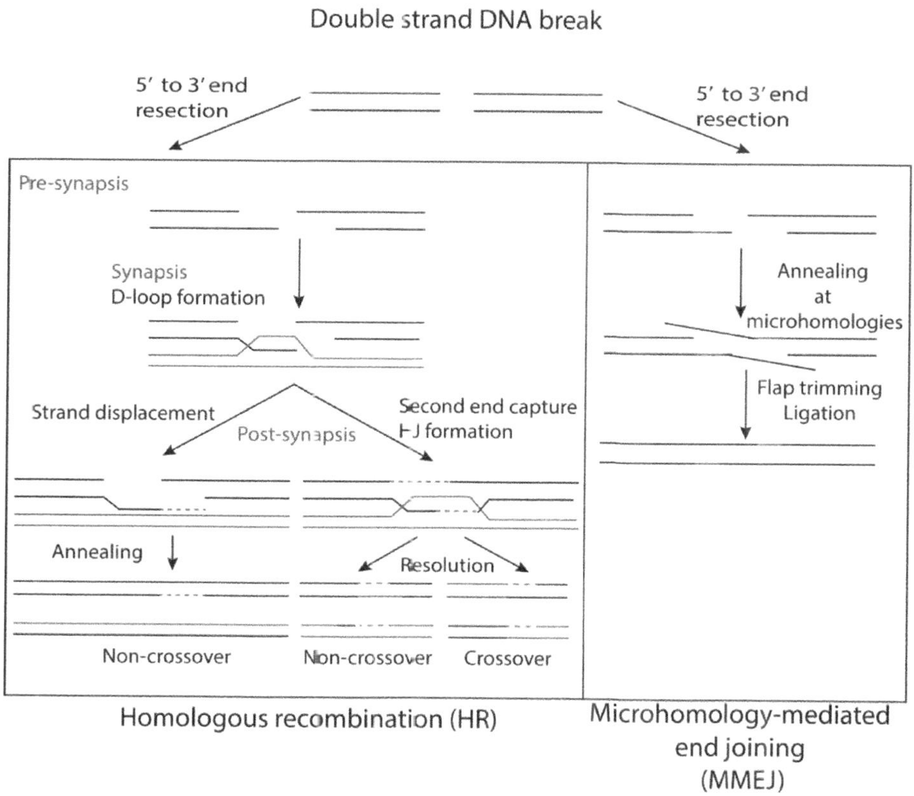

Figure 2.7. *In archaea, two major pathways repairing DSBs are homologous recombination (HR) and microhomology-mediated end joining (MMEJ). For a detailed discussion, see the text. For a color version of this figure, see www.iste.co.uk/clouet/archaea2.zip*

MMEJ has been described in halophiles and *Sulfolobales*. Following DSB, microhomologies are revealed by cellular exonucleases, allowing complementary sequences to anneal. This produces a flapped structure. The single-stranded extremities are likely cleaved by Fen1 endonuclease or the RecJ/GINS-associated nuclease (GAN). Then, nicks are ligated by DNA ligase to complete DNA repair. In the case where the microhomologies are located far from the site of damage, often the intervening sequences are lost during this repair process. NHEJ does not require regions of homology. Instead, broken ends are brought together in a protein-mediated complex. Although the molecular details of archaeal NHEJ have not been determined, it is likely that the DNA end-binding Ku protein, conserved in some archaeal species, bounds the DNA extremities to allow for the subsequent processing of the DNA ends to allow gap filling and ligation.

2.4. Coordination of the different pathways by the replication clamp

Many different types of DNA damage either result from errors of DNA replication or halt the replisome. In addition, some archaeal DNA repair proteins such as Fen1 and potentially DNA polymerase PolB are also multitasking between DNA replication and repair. These observations suggest that efficient and specific mechanisms must exist that allow for the coordination of DNA replication with the various DNA repair mechanisms outlined above.

In all domains of life, the replication (sliding) clamp acts as a major conductor of the different DNA transactions. In archaea, both homo- and heterotrimeric replication clamps with a ring-shaped three-dimensional structure have been described (MacNeill 2016). These proteins were originally named “proliferating cell nuclear antigen (replication clamp PCNA)” as their human ortholog was originally identified as an antigen present in the nuclei of dividing cells that reacted with an autoantibody in the sera of patients with lupus erythematosus. Later, it was demonstrated that human PCNA encircles DNA and acts as a processivity factor of eukaryotic replicative DNA polymerases.

PCNA structures from a range of archaeal species have been reported. These include homotrimers from the thermophilic euryarchaeal species *Pyrococcus furiosus* and *Archaeoglobus fulgidus* and the mesophilic haloarchaeon *Haloferax volcanii*. *Sulfolobus* species and other *crenarcheota* use the PCNA heterotrimer (Pan et al. 2011). The three-dimensional structures of archaeal PCNA are practically identical to their human counterpart. The inner diameter of the PCNA ring is approximate 3.5 nm, which is sufficiently large to accommodate the DNA helix (B-form) with a diameter of 2.0 nm (Figure 2.8). One face of the PCNA ring interacts with many DNA replication and repair proteins, whereas the other face is not implicated in direct protein–protein interactions. The outer surface of the ring-shaped PCNA trimer carries negative charges, whereas the central cavity accommodating DNA is positively charged. One PCNA monomer consists of two domains that are linked by a domain-connecting loop. This loop is evolutionary conserved and mediates specific interactions of the replication clamp with its interaction partners.

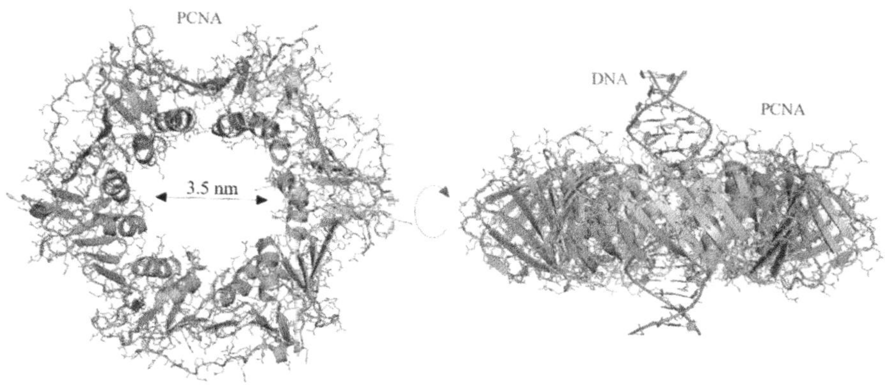

Figure 2.8. *Two orthogonal views of the archaeal PCNA in complex with DNA. The monomers of the homotrimeric structure of PCNA are colored in cyan, green, and violet. The diameter of the pore in PCNA is indicated by an arrow. For a color version of this figure, see www.iste.co.uk/clouet/archaea2.zip*

To be functionally active, different PCNA-interacting proteins bind and dissociate in a temporally ordered manner. It is somewhat surprising that the identified peptide domain motifs of archaeal PCNA-interacting proteins (PIP-motifs) are quite similar and have overlapping binding sites with PCNA (Table 2.1). For instance, the sequence motif Q-x(2)-(h)-x(2)-(a)-(a) (h, moderately hydrophobic residue; a, a highly hydrophobic residue) is known to mediate interactions of archaeal and human PCNA with many partners (see, for example, Meslet-Cladiere et al. 2007). The sequence variations of the archaeal PCNA-binding motifs may underpin differential binding of the different protein partners of the PCNA, thus facilitating the coordination of the different DNA transactions.

Many examples of the functional interactions between PCNA and other proteins have been described in the literature. For instance, PCNA markedly stimulates archaeal DNA polymerases, DNA ligases and endonucleases. Moreover, the interaction between PCNA and UDG participating in archaeal BER has been reported in both *crenarchaeota* and *euryarchaeota* and it has been proposed that such interaction allows UDGs to be recruited to the replication fork once DNA polymerase is blocked at the uracil. Thus, the archaeal PCNA likely functions as a platform that allows coordination of the different DNA repair pathways. Table 2.2 indicates additional examples of the archaeal PCNA interactions.

Archaeal protein	Effect of the PCNA	DNA repair pathway	Organisms (examples)
Fen1	Stimulation	BER, RER	*P. abyssi, S. solfataricus*
RnaseH2	Recruitment	RER	*P. abyssi, A. fulgidus*
PolB	Increased processivity	Many	*P. furiosus, S. solfataricus*
UDG	Stimulation	BER	*P. aerophilum, P. furiosus*
DNA ligase	Recruitment	Many	*P. abyssi*
AP endonuclease	Stimulation	BER	*P. furiosus*
XPF	Stimulation	NER?	*S. solfataricus*
NucS/EndoMS	Stimulation	MMR	*Thermococcales*
Hjc	Stimulation (nuclease)	HR	*S. solfataricus*
Hjm	Stimulates (helicase)	HR	*P. furiosus*
Mre11-Rad50	Modulation of nuclease specificity	HR	*P. furiosus*

Table 2.2. *Examples of functional modulation of archaeal DNA repair enzymes by the replication clamp PCNA*

2.5. Summary and conclusion

The extensive efforts from comparative genomics, genetics, biochemistry and structural biology during the past years have revealed that archaea use many parallel ways to guarantee the maintenance of genetic information sometimes under very harsh conditions. The goal of this chapter on archaeal DNA repair was to highlight how endogeneous and exogeneous factors cause DNA damage and how this can be related to environmental factors such as high temperature. We have also underlined the phylogenetic (Table 2.3) and mechanistic diversity of how the damaged DNA is efficiently repaired by multiple archaeal DNA repair pathways. Strikingly, it seems that archaea can not only use unique DNA repair enzymes but also their "bacterial" and "eukaryotic" counterparts for DNA repair. We imagine that during the next years we will continue to discover new archaeal DNA repair enzymes in the so-far understudied novel phylogenetic groups of archaea that have only recently been discovered (or are still being discovered).

Group	RPA	EndoMS (NucS)	Uvr ABC	Mut SL	XPF euk/short		XPBB	XPD	Bax1
TACK									
Sulfolobus									
Cenarchaeum									
Thorarchaeum									
Lokiarcheum									
Euryarchaea									
Thermococcus									
Methanosarcina									
Haloferax									
Thermoplasmama									
Nanoarchaeum									

Table 2.3. *Simplified presentation of the phylogenetic distribution of the major DNA repair genes in key archaeal species from TACK and euryarcheota lineages. For a color version of this table, see www.iste.co.uk/clouet/archaea2.zip*

2.6. References

Cann, I.K., Komori, K., Toh, H., Kanai, S., Ishino, Y. (1998). A heterodimeric DNA polymerase: Evidence that members of Euryarchaeota possess a distinct DNA polymerase. *Proc. Natl. Acad. Sci. USA*, 95(24), 14250–14255.

Castañeda-García, A., Prieto, A.I., Rodríguez-Beltrán, J. Alonso, N., Cantillon, D., Costas, C., Pérez-Lago, L., Zegeye, E.D., Herranz, M., Plociński, P. et al. (2017). A non-canonical mismatch repair pathway in prokaryotes. *Nature Communications*, 8, 14246.

Cooper, C.D.O. (2018). Archaeal DNA polymerases: New frontiers in DNA replication and repair. *Emerg. Top Life Sci.*, 2(4), 503–516.

Gehring, A.M., Zatopek, K.M., Burkhart, B.W., Potapov, V., Santangelo, T.J., Gardner, A.F. (2020). Biochemical reconstitution and genetic characterization of the major oxidative damage base excision DNA repair pathway in Thermococcus kodakarensis. *DNA Repair (Amst)*, 86, 102767.

Grasso, S. and Tell, G. (2014). Base excision repair in archaea: Back to the future in DNA repair. *DNA Repair (Amst)*, 21, 148–157.

Heider, M.R., Burkhart, B.W., Santangelo, T.J., Gardner, A.F. (2017). Defining the RNaseH2 enzyme-initiated ribonucleotide excision repair pathway in archaea. *J. Biol. Chem.*, 292(21), 8835–8845.

Ishino, S., Makita, N., Shiraishi, M., Yamagami, T., Ishino, Y. (2015). EndoQ and EndoV work individually for damaged DNA base repair in *Pyrococcus furiosus*. *Biochimie*, 118, 264–269.

Ishino, S., Skouloubris, S., Kudo, H., l'Hermitte-Stead, C., Es-Sadik, A., Lambry, J.C., Ishino, Y., Myllykallio, H. (2018). Activation of the mismatch-specific endonuclease EndoMS/NucS by the replication clamp is required for high fidelity DNA replication. *Nucleic Acids Res.*, 46(12), 6206–6217.

Kunkel, T.A. and Erie, D.A. (2015). Eukaryotic mismatch repair in relation to DNA replication. *Annu. Rev. Genet.*, 49, 291–313.

Lindahl, T. (1993). Instability and decay of the primary structure of DNA. *Nature*, 362, 709–715.

MacNeill, S.A. (2016). PCNA-binding proteins in the archaea: Novel functionality beyond the conserved core. *Curr. Genet.*, 62, 527–532.

Makarova, K.S., Sorokin, A.V., Novichkov, P.S., Wolf, Y.I., Koonin, E.V. (2007). Clusters of orthologous genes for 41 archaeal genomes and implications for evolutionary genomics of archaea. *Biol. Direct.*, 2(33). doi: 10.1186/1745-6150-2-33.

Meslet-Cladiére, L., Norais, C., Kuhn, J., Briffotaux, J., Sloostra, J.W., Ferrari, E., Hübscher, U., Flament, D., Myllykallio, H. (2007). A novel proteomic approach identifies new interaction partners for proliferating cell nuclear antigen. *J. Mol. Biol.*, 372(5), 1137–1148.

Minobe, A., Fukui, K., Yonezu, H., Ohshita, K., Mizobuchi, S., Morisawa, T., Hakumai, Y., Yano, T., Ashiuchi, M., Wakamatsu, T. (2019). Biochemical characterization of mismatch-binding protein MutS1 and nicking endonuclease MutL from a euryarchaeon *Methanosaeta thermophila*. *DNA Repair (Amst)*, 75, 29–38.

Nakae, S., Hijikata, A., Tsuji, T., Yonezawa, K., Kouyama, K.I., Mayanagi, K., Ishino, S., Ishino, Y., Shirai, T. (2016). Structure of the EndoMS-DNA complex as mismatch restriction endonuclease. *Structure*, 24(11), 1960–1971.

Pan, M., Kelman, L.M., Kelman, Z. (2011). The archaeal PCNA proteins. *Biochem. Soc. Trans.*, 39(1), 20–24.

Pérez-Arnaiz, P., Dattani, A., Smith, V., Allers, T. (2020). Haloferax volcanii – A model archaeon for studying DNA replication and repair. *Open Biol.*, 10(12), 200293.

Rouillon, C. and White, M.F. (2011). The evolution and mechanisms of nucleotide excision repair proteins. *Res. Microbiol.*, 162(1), 19–26.

Walker, J.E., Luyties, O., Santangelo, T.J. (2017). Factor-dependent archaeal transcription termination. *Proc. Natl. Acad. Sci. USA*, 114(33), E6767–E6773.

White, M.F. and Allers, T. (2018). DNA repair in the archaea – An emerging picture. *FEMS Microbiol. Rev.*, 42(4), 514–526.

3

Transcription in Archaea

Thomas FOUQUEAU[1], Duy KHANH PHUNG[1] and Ludovic SAUGUET[2]

[1]RNAP, ISMB, University of London (UCL), United Kingdom

[2]Institut Pasteur, Université Paris Cité, France

In all living beings, transcription is the first step in gene expression. Transcription consists of copying coding DNA sequences, known as "genes", and transcribing them into RNA molecules (Figure 3.1). While the DNA molecule is the universal medium for genetic information, the messenger RNA molecule is the medium recognized by ribosome, which translates the genetic information into protein sequences. The enzyme that catalyzes this transcription reaction is called RNA polymerase (RNAP). During transcription, the RNAP separates double-stranded DNA and adds ribonucleotide triphosphates (rNTPs) to the 3′ end of the RNA chain through synergy with the template DNA strand, the strand transcribed by the enzyme.

Transcription in archaea displays both bacterial and eukaryotic characteristics. Archaea have a compact genome, and many archaeal genes are organized in operons, while many messenger RNAs are polycistronic, as in bacteria. In addition, gene expression is regulated by bacterial regulators that bind to the gene promoter. In contrast, the transcription machinery of archaea is evolutionarily closer to that of eukaryotes in terms of (i) the organization of gene promoter sequences, (ii) the composition of the various RNAP subunits and (iii) the transcription factors involved. The archaeal transcription mechanism can thus be considered a simplified version of that of eukaryotes (Werner and Grohmann 2011).

Most of the model organisms used to study the molecular biology of archaea are extremophilic, and particularly thermophilic. Proteins from thermophilic organisms

Biology of Archaea 2,
coordinated by Béatrice CLOUET-D'ORVAL, Bruno FRANZETTI and Philippe OGER.

are generally more stable than their mesophilic counterparts, which is an advantage for recombinant expression, in vitro enzymatic assays and structural studies. Over the last 20 years, several biochemical tools have been developed to study the function of archaeal transcription machinery in vitro. In particular, the ability to reconstitute RNAP and transcription complexes in vitro, from recombinant proteins, has enabled biochemical and biophysical studies that were previously impossible in bacterial and eukaryotic systems (Werner and Weinzierl 2002; Naji et al. 2007). These studies have led to the functional characterization of the archaeal RNAP subunits and their factors, improving overall understanding of the transcription mechanism in all three domains of life. In addition, recent advances in genetic tools, combined with high-throughput sequencing, are now providing valuable information on the mechanism and regulation of RNA synthesis in vivo. The transcription process in archaea is therefore an ideal model for studying the fundamental mechanisms of transcription, and for gaining a better understanding of the evolution of the transcription machinery across the three domains of life.

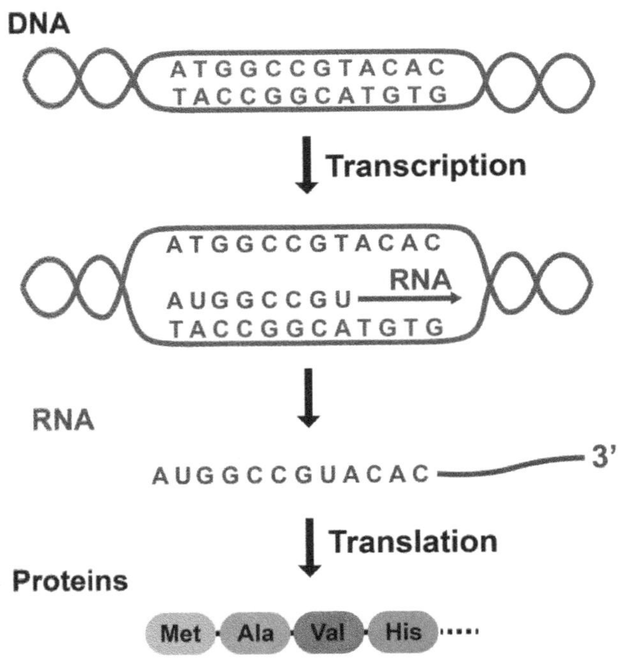

Figure 3.1. *Fundamental theory of molecular biology. DNA, the stable, transmissible carrier of genetic information, is transcribed into RNA, which can then be translated into proteins. For a color version of this figure, see www.iste.co.uk/clouet/archaea2.zip*

3.1. RNA polymerase

3.1.1. *RNA polymerase architecture in archaea*

	Bacteria	Archaea (Eury-/DPANN; TACK/ASGARD)	Eukaryotes (RNAPII)
Main subunits	β'	Rpo1	RPB1
	β	Rpo2	RPB2
	α_{I}	Rpo3	RPB3
	α_{II}	Rpo11	RPB11
	ω	Rpo6	RPB6
Accessory subunits		Rpo5	RPB5
		Rpo10	RPB10
		Rpo12	RPB12
		Rpo4	RPB4
		Rpo7	RPB7
		Rpo8* (TACK/ASGARD)	RPB8
			RPB9
		Rpo13* (TACK/ASGARD)	
Initiation factors	σ^{70}	TBP	TBP
		TFB	TFIIB
		TFEα	TFIIEα
		TFEβ*	TFIIEβ
Elongation factors	Gre	TFS	TFIIS
		Spt4	Spt4
	NusG	Spt5	Spt5
		Elf1* (TACK/ASGARD)	Elf1
	NusA	NusA	

Table 3.1. *Conservation of RNA polymerase subunits and general transcription factors. For a color version of this table, see www.iste.co.uk/clouet/archaea2.zip*

All of these RNAPs share a highly conserved active site, composed of five subunits that contain all of the elements required for transcription (Cramer 2002; Werner and Grohmann 2011; Fouqueau et al. 2018).

These subunits form an active site which, metaphorically, is often compared to a crab claw, whose claw tips would interact with the template DNA duplex (see the following paragraph) (Cramer et al. 2001; Lane and Darst 2010a, 2010b).

The two main subunits that form the core of the enzyme – which are Rpo1 and Rpo2 in archaeal RNAP, Rpb1 and Rpb2 in eukaryotic RNAPII, and β and β′ in bacterial RNA – host the active polymerase site, whose catalytic residues have been mapped by numerous biochemical and genetic studies. The association between these two catalytic subunits is stabilized by three assembly subunits (Rpo3, Rpo11 and Rpo6 in archaea).

In archaea and eukaryotes, but not in bacteria, several auxiliary subunits associate with the catalytic core subunits and play an important role in RNA polymerase assembly and stability (Rpo10 and 12), interaction with DNA (Rpo5 and 13), neo-synthesized RNA, as well as the TFE initiation factor (Rpo4 and 7).

In addition, RNAP activity is regulated by numerous transcription factors during the three main phases of the transcription cycle: initiation, elongation and termination. The functions of the subunits and the main transcription factors that make up RNAPs in the three domains of life are summarized in Table 3.1 and Figure 3.2.

Table 3.1 shows the conservation of RNA polymerase subunits and transcription initiation and elongation factors in bacteria, archaea and eukaryotes. Within the archaea, the transcription machinery of species belonging to the TACK and ASGARD superphyla is more closely related to the eukaryotic RNAPII system than species belonging to the *Superphyla Euryarchaeota* and DPANN. Subunits and factors that are only conserved in certain species are indicated with a star.

Functionally analogous factors are indicated by dotted lines. The bacterial sigma-70 factor is functionally analogous to the TBP/TFB duo, although it only shares a very limited sequence similarity with TFB. Bacterial Gre factors are functionally analogous to TFS/TFIIS factors but have a different evolutionary origin.

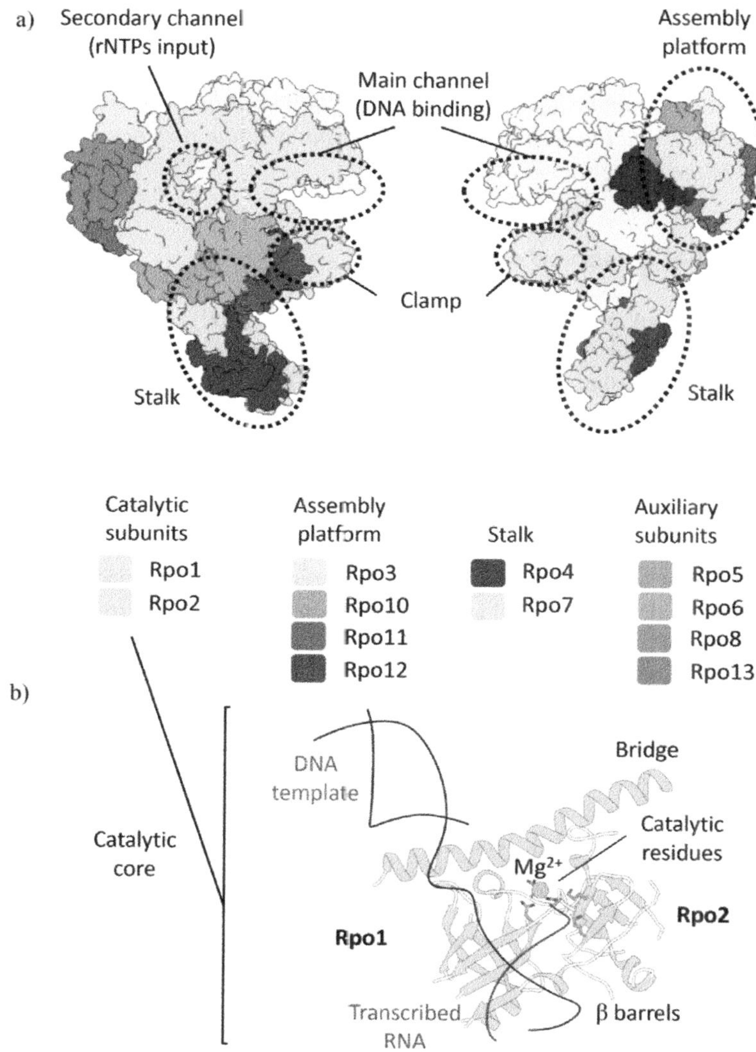

Figure 3.2. *RNAP structure. For a color version of this figure, see www.iste.co.uk/clouet/archaea2.zip*

COMMENT ON FIGURE 3.2.– *(a) It represents the overall architecture of archaeal RNAP, with the different subunits differentiated according to a color code defined in the central panel. The structure shown is the crystallographic structure of* Sulfolobus shibatae *RNAP (Korkhin et al. 2009); (b) it shows the structure of the catalytic core of* Saccharomyces cerevisiae *RNAP II (Wang et al. 2006) (no archaeal RNAP structure has yet been determined in complex with DNA). The catalytic core is located at the interface between two β barrels. The first β barrel carries a triad of canonical aspartate residues that coordinate one of the catalytic magnesium ions, and the second β barrel contributes lysine residues that help guide substrates in the active site.*

3.1.2. *A highly preserved catalytic core*

Like all known polymerases, RNA polymerases catalyze the formation of a phosphodiester bond between two nucleotides using a two-metal catalytic mechanism (Steitz 1998). During catalysis, the first metal acts as a Lewis acid, lowering the pKa of the primer's 3′-OH group and activating its nucleophilic attack on the α phosphate of the incoming nucleotide. The second metal helps stabilize the pentavalent transition state formed around α phosphate and facilitates the departure of the pyrophosphate group by neutralizing its negative charges. The two catalytic metals are generally magnesium ions, coordinated by the side chains of two or three aspartate residues. In RNAPs, the catalytic residues are carried by a highly conserved catalytic core, located at the interface between two β barrels (Figure 3.2). The first β barrel carries a triad of canonical aspartate residues that coordinate one of the catalytic magnesium ions, and the second β barrel contributes lysine residues that help orient substrates in the active site. Recently, structural studies and bioinformatics predictions have revealed that this catalytic core is present in a wide variety of RNA and DNA polymerases, which participate in transcription, DNA replication and the regulation of gene expression (Iyer et al. 2003; Salgado et al. 2006; Sauguet et al. 2016). These studies thus show that the existence of the catalytic core of two β barrel RNAPs appeared very early in evolution, probably predating the existence of the last common ancestor and diversification between the three domains of life (Fouqueau et al. 2017a; Sauguet 2019).

3.1.3. *Anatomy of the active site of an RNA polymerase*

The two β barrel catalytic core occupies a central position in the RNAP active site, located between three channels: (i) the main channel, through which template DNA is projected toward the catalytic core; (ii) the secondary channel, which supplies the catalytic core with rNTPs; (iii) the exit channel, which allows the neo-synthesized RNA to escape (Markovtsov et al. 1996; Zaychikov et al. 1996; Nudler 1999; Lane and Darst 2010a, 2010b). The insertion of rNTPs and the translocation of RNAP along the DNA are coordinated through mobile elements known as trigger-loops and bridge-helixes. The trigger-loop is a highly flexible polypeptide loop and the bridge-helix is an α-helix capable of bending at its center, functioning as a ratchet during RNAP translocation. Other highly conserved structural elements help maintain the structure of the transcription bubble: the fork loop 2, which separates the template DNA strand from the complementary strand; the rudder, which stabilizes the DNA–RNA hybrid in the active site; the lid, which regulates the extension of the DNA–RNA hybrid and facilitates dissociation of the neo-synthesized RNA from the template DNA; the switch 3 loop, which guides neo-synthetized RNA toward the "RNA exit channel"; and the zipper, which helps reform double-stranded DNA upstream of the transcription bubble. The role of these different structural elements in the elongation process will be explained in section 3.2.2.

3.2. The three stages of transcription

In all living organisms, transcription takes place in three distinct stages: initiation, elongation and termination (Figure 3.3).

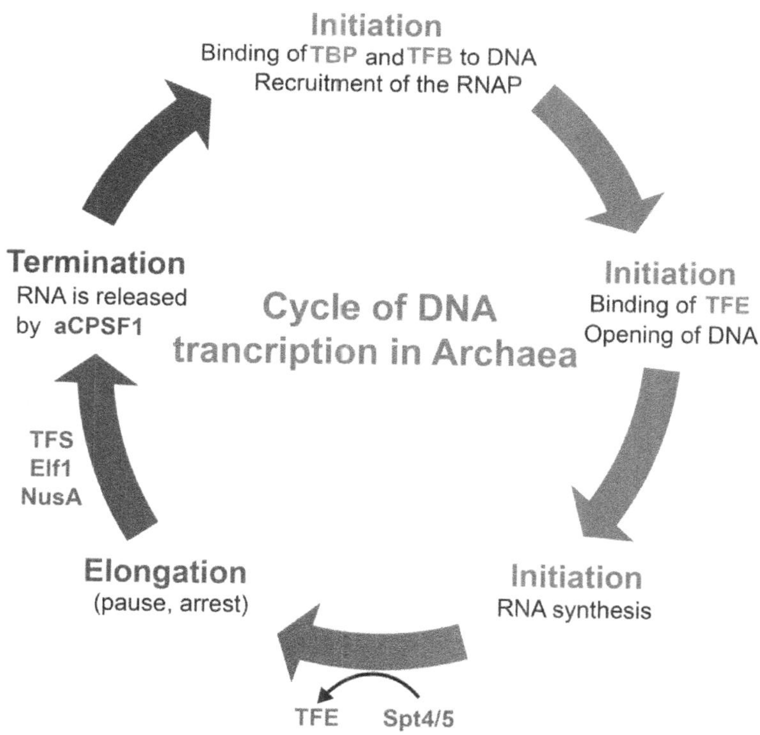

Figure 3.3. *Transcription cycle in archaea. For a color version of this figure, see www.iste.co.uk/clouet/archaea2.zip*

COMMENT ON FIGURE 3.3.– *The transcription cycle in archaea comprises initiation, elongation and termination phases, during which RNAP is assisted by general transcription factors. During the transition from initiation to elongation, RNA polymerase-bound initiation factors (blue) are replaced by elongation factors (green). At termination, the RNAP dissociates from the DNA, releasing the synthesized RNA, starting a new transcription cycle.*

3.2.1. *Transcription initiation*

RNAP alone is not capable of initiating transcription. In fact, initiation is controlled by general initiation factors, which form a complex capable of recruiting RNAP to the promoter. Archaeal RNAP uses three general initiation factors, called TBP (*TA*TA-box *b*inding *p*rotein), TFB (*t*ranscription *f*actor *B*) and TFE (*t*ranscription *f*actor *E*), respectively, similar to the TBP, TFIIB and TFIIE factors of eukaryotes (Table 3.1) (Hausner et al. 1996; Bell et al. 2001a; Hanzelka et al. 2001; Blombach et al. 2015). The association between RNAP and initiation factors forms a PIC (*p*re-*i*nitiation *c*omplex) that is capable of initiating transcription.

3.2.1.1. *The promoter*

A promoter is a DNA region that is located upstream of a gene or operon, which is essential for transcription initiation. The organization of promoter sequences in archaea is very similar to that of eukaryotes (Reiter et al. 1998). In archaea, the promoter essentially consists of two building blocks, TATA-box and BRE, which direct the assembly of transcription initiation complexes. These elements enable the TBP and TFB initiation factors to be recruited to the promoter. The TATA-box is a conserved sequence located between 25 and 30 base pairs (bp) upstream of the transcription start site (TSS; position + 1) (Thomm and Wich 1988; Rowlands et al. 1994; Bartlett et al. 2000). Consisting of 7–8 bp and rich in A and T nucleotides, the TATA-box enables recruitment of the initiation factor TBP. The second element, called BRE (B recognition element), is located upstream of the TATA-box and participates in the recruitment of the initiation factor TFB (Bell et al. 1999a). A third element, only composed of a dinucleotide, and called Initiator (Inr), is also conserved in archaeal promoters. The archaeal Inr is composed of a T-A/G dinucleotide, at a position of -1 and +1, respectively, which manages the direction of the transcription site in vivo (Smollett et al. 2017).

3.2.1.2. *The initiation complex*

Transcription initiation by RNAP begins with PIC assembly at the promoter (Figure 3.4). Archaeal PIC formation begins with TBP binding to the TATA-box. The archaeal TBP factor is composed of two symmetrically organized DNA-binding domains, which looks like a "horse saddle" (Patikoglou et al. 1999). When TBP binds to the TATA-box, it bends the DNA ~90° by inserting phenylalanine residues between the base pairs surrounding the TATA-box DNA sequence (Bewley et al. 1998). DNA deformation then increases the contact surface between DNA and TBP, thereby improving the stability of the TBP-DNA complex (Whittington et al. 2008).

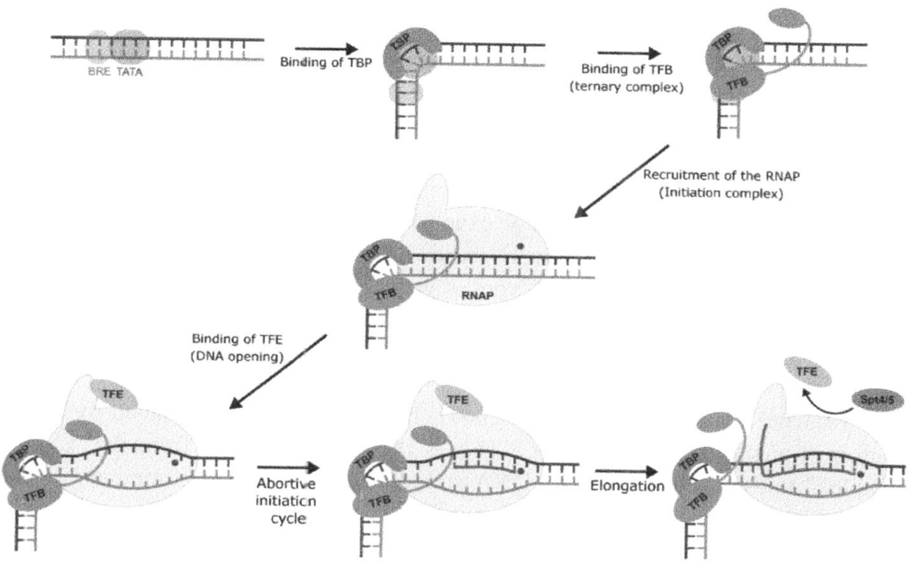

Figure 3.4. *Transcription initiation. For a color version of this figure, see www.iste.co.uk/clouet/archaea2.zip*

COMMENT ON FIGURE 3.4.– *Transcription initiation begins with the recruitment of the initiation factors TBP (pink) and TFB (green) to the TATA-box and BRE promoter elements, forming a ternary complex. The RNA polymerase (RNAP) (gray) is then recruited to form the minimum PIC. TFE (yellow) is recruited to the initiation complex and stimulates DNA strand separation and formation of the transcription bubble, enhancing the transition from a closed to an open complex. In the presence of NTP substrates, RNAP enters an abortive initiation cycle, during which RNAs of 3–9 nt are synthesized. During the transition from initiation to elongation, the elongation factor Spt4/5 (orange) displaces TFE in a factor exchange process. Template and non-template DNA strands are shown in dark and light blue, respectively. Catalytic Mg-A is shown in purple.*

In most archaeal species, the formation of a stable TBP–DNA complex also requires the recruitment of the TFB factor (Gietl et al. 2014). TFB binds to one side of TBP to be in contact with the BRE motif of the promoter (Kosa et al. 1997). TFB is composed of three domains: an N-terminal domain with a zinc finger motif; a C-terminal domain composed of a repeated cyclin-fold and a central, flexible domain that connects the previous two domains. The central domain contains B-reader and B-linker motifs (Kostrewa et al. 2009). TFB binds to the BRE motif of the promoter through its second *cyclin* motif. The position of BRE relative to the TATA-box determines the orientation of the DNA–TBP–TFB complex, as well as the direction of transcription (Nikolov et al. 1995; Bell et al. 1999a). The zinc finger domain of TFB interacts with the dock domain of RNAP and recruits the enzyme to the promoter to form the PIC (Kostrewa et al. 2009). Most archaeal species possess several TBP and TFB homologs. For example, *Halobacterium* species possess six TBPs and seven TFBs, respectively, whose different combinations specifically enable the transcription of different types of genes (Baliga et al. 2000; Facciotti et al. 2007). During PIC formation, the B-reader and B-linker motifs of the TFB penetrate deep into the active center of the RNAP, stimulating DNA strand separation and stabilization of the template DNA strand in the active center of the RNAP (Werner and Weinzierl 2005; Kostrewa et al. 2009). This process is accompanied by major conformational changes in the PIC. This is the transition stage from the "closed complex" to the "open complex". DNA strand separation and the formation of the "open" PIC are facilitated by the third archaeal initiation factor, TFE.

The TFE factor is a heterodimeric complex composed of two subunits, called TFEα and TFEβ. The TFEα subunit consists of an N-terminal winged-helix domain and a C-terminal zinc finger domain (Bell et al. 2001b; Meinhart et al. 2003). TFEα interacts with the clamp coiled-coil and stalk domains of RNAP, retaining the RNAP clamp domain in an open configuration (Schulz et al. 2016). Once bound to RNAP, TFEα interacts with the non-transcribed DNA strand, facilitating transition between the "closed" and "open" PIC (Grohmann et al. 2011). The TFEβ subunit is not conserved in all archaeal species. It is absent in most species of the phylum *Euryarchaea*, but conserved in species of the phylum TACK (with the exception of *Korarchaea*) and ASGARD (Blombach et al. 2016). In these species, TFEβ significantly stimulates the TFEα function. The TFEβ subunit contains an N-terminal winged-helix domain in addition to a C-terminal (4Fe-4S) cluster domain (Blombach et al. 2015). The (4Fe-4S) cluster of TFEβ is sensitive to nutrient deficiency and external stress, such as oxidation (Blombach et al. 2015; Bischof et al. 2018). Under these conditions, the amount of TFEβ decreases in vivo, and the expression of around a third of genes is then significantly affected (Blombach et al. 2015, 2018). Thus, in addition to its role as an initiation factor, TFEβ also acts as a global regulator of transcription.

3 2.1.3. *The abortive initiation cycle*

Once the formation of the “open” PIC is complete, RNAP can begin transcribing DNA (Figure 3.4). After DNA opening, the positioning of the template strand in the active site and the arrival of rNTPs, via the “secondary channel”, induce the formation of the first phosphodiester bond between two rNTPs in addition to the template strand. The RNAP then enters a phase of non-productive transcription known as the “abortive initiation cycle”. The region downstream of the transcription initiation site is repeatedly transcribed by RNAP, which synthesizes small RNAs (3–12 nucleotides (nt)) without detaching from the template strand (Revyakin et al. 2006). Transcription cannot proceed to the elongation phase until the contact between RNAP and TFB is broken. Thus, when emerging RNA exceeds 5 nt, it collides with the B-reader and B-linker domains of the TFB, disrupting the interaction between the TFB and the active center of the RNAP (Sainsbury et al. 2013). Once the RNA length exceeds 15 nt, the contact between RNAP and TFB is broken, triggering the passage of RNAP into a phase of productive synthesis, known as the elongation phase (Dexl et al. 2018).

3.2.2. *Elongation*

During elongation, RNAP advances along the DNA and adds additional rNTPs to the template DNA strand, at the 3′ end of the neo-synthesized RNA strand (Figure 3.5). In the active center of the RNAP, a DNA–RNA heteroduplex is formed between the transcribed RNA and the template DNA strand. As the RNAP progresses, the DNA double helix opens downstream of the polymerase and closes upstream of the polymerase, forming the transcription “bubble”. The size of the transcription bubble, as well as the size of the DNA–RNA hybrid, are controlled by loops in the active center of the RNAP. Once separated from the template DNA strand, the 5′ end of the neo-synthesized RNA exits the RNAP through a narrow channel, known as the “RNA exit channel”, before binding to the S1 motif of the RNAP stalk. It has been shown in vitro that the interaction between RNA and the stalk regulates the efficiency of transcription elongation and termination, by increasing the processivity of RNAP (Hirtreiter et al. 2010a; Grohmann and Werner 2011).

During the transition from initiation to elongation (commonly known as promoter escape), initiation factors are exchanged for elongation factors, which bind to RNAP to form the transcription elongation complex (TEC). These factors increase the processivity and stability of RNAP, by attenuating transcriptional pauses and stops, enabling the efficient transcription of long genes.

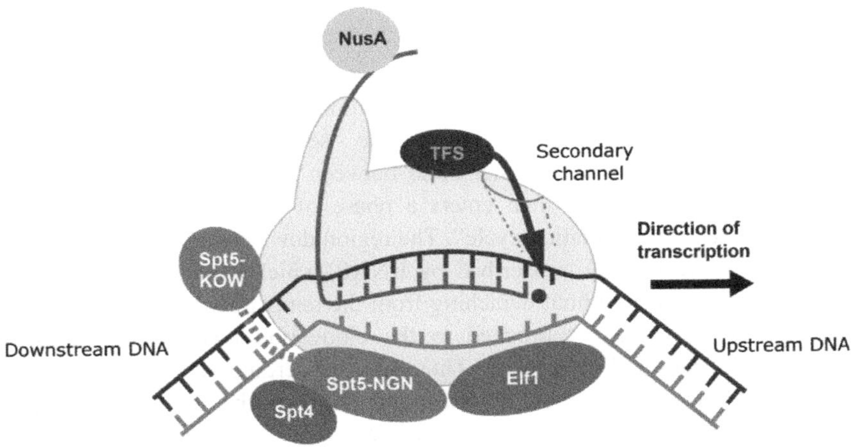

Figure 3.5. *Schematic representation of the archaeal transcription elongation complex. The complex comprises the RNAP bound to DNA and RNA, and the elongation factors TFS, Spt4/5 and Elf1. NusA is the only elongation factor that does not bind directly to RNA polymerase, but to synthesized RNA. For a color version of this figure, see www.iste.co.uk/clouet/archaea2.zip*

3.2.2.1. *The nucleotide addition cycle*

During the elongation phase (Figure 3.5), the TEC moves along the DNA in a Brownian ratchet motion. The energy required for the complex to advance is provided by the incorporation of rNTPs, complementary to the template DNA strand (Bar-Nahum et al. 2005). Thus, the elongation phase consists of a cycle of complementary rNTP additions to the template DNA strand. An 8–9 bp DNA–RNA hybrid is formed in the active center of the RNAP, inside the "main channel" formed by the catalytic subunits (Rpo1 and Rpo2) (Kireeva et al. 2000; Vassylyev et al. 2007a). The 3′ end of the RNA (known as the i site) is close to the active site formed by a magnesium ion. During the nucleotide addition cycle, the rNTP substrate, which coordinates a second Mg^{2+} ion, enters the active center of RNAP through the "secondary channel", which is orthogonal to the main channel. The substrate is positioned in the insertion site (known as the i + 1 site), located at the 3′ end of the 8-bp DNA–RNA hybrid (Vassylyev et al. 2007b). A nucleophilic reaction then occurs between the 3′ end of the RNA strand and the α-phosphate of the incoming rNTP, forming a phosphodiester bond, followed by release of the pyrophosphate (Ppi) and the translocation of the TEC by one base pair along the DNA. This translocation moves the new 3′ end of the RNA from the i + 1 site to the i site, making way for the next substrate. The nucleotide addition cycle is thus divided into two steps: (i) rNTP insertion, followed by (ii) translocation.

The active center of RNAP (Figure 3.6) contains two mobile elements, called trigger-loop and bridge-helix, which coordinate the insertion of complementary rNTPs and the translocation of RNAP along the DNA, respectively. As a reminder, the trigger-loop is a highly flexible polypeptide loop and the bridge-helix is an α-helix that is capable of bending at its center, functioning as a ratchet during RNAP translocation (Wang et al 2006; Brueckner and Cramer 2008). During the nucleotide addition cycle, the trigger-loop interacts with the incoming rNTP and selects the rNTP that is complementary to the template DNA strand, through a network of interactions between the trigger-loop and all parts of the rNTP (base, ribose and phosphates) (Wang et al. 2006; Kaplan et al. 2008; Yuzenkova et al. 2010; Fouqueau et al. 2013). Once the rNTP has been recognized, the trigger-loop then catalyzes rNTP incorporation at the 3′ end of the RNA strand. Incorporation of the rNTP is followed by a conformational change in the bridge-helix, that displaces the RNAP by 1 bp relative to the DNA–RNA hybrid (Tuske et al. 2005; Weinzierl 2010). This translocation positions the new 3′ end of the RNA strand at the i site, enabling the insertion of a new rNTP at the i + 1 site. The crucial role of these two elements during the nucleotide addition cycle has been studied by directed mutagenesis experiments in archaea that have produced low-fidelity or "super active" RNAP variants (Tan et al. 2008; Fouqueau et al. 2013).

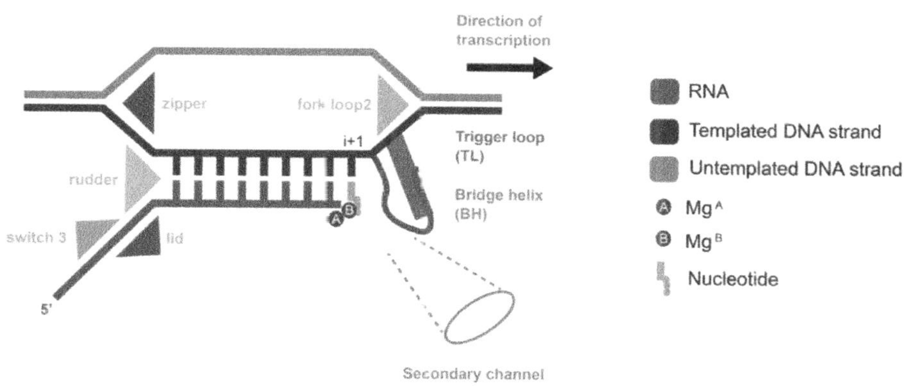

Figure 3.6. *Structural elements essential for maintaining DNA–RNA hybrid integrity and RNA extension. Schematic representation of the active center of RNA polymerase during the elongation phase. The NTP insertion site corresponds to the i + 1 site. Structural elements essential for maintaining the RNA–DNA hybrid are represented by colored triangles, and the trigger loop (TL) and bridge helix (BH) are shown in pink and green, respectively. For a color version of this figure, see www.iste.co.uk/clouet/archaea2.zip*

3.2.2.2. *Structural elements maintaining the integrity of the DNA–RNA hybrid*

In addition to these mobile structural elements, the active center of RNAP (Figure 3.6) contains highly conserved polypeptide loops that form an interaction network with nucleic acids to increase TEC stability. Directed mutagenesis studies have demonstrated that interactions between these loops and nucleic acids are essential for maintaining the correct size of the transcription bubble and the DNA–RNA hybrid. During elongation, double-stranded DNA collides with the RNAP fork loop 2, which separates the two DNA strands and prevents re-association of the DNA strands, so as to maintain the transcription bubble (Vassylyev et al. 2007a; Naji et al. 2008). The DNA template strand then finds its way into the active center of RNAP and, as transcription proceeds, forms an 8–9 bp hybrid with the transcribed RNA. The rudder loop stabilizes the hybrid in the active center by interacting with the 5′ end of the hybrid (Kuznedelov et al. 2002). Hybrid size is controlled by the lid loop, which sterically blocks hybrid extension beyond 8–9 bp and facilitates dissociation of neo-synthesized RNA and template DNA (Naryshkina et al. 2006; Toulokhonov and Landick 2006; Naji et al. 2008). Switch loop 3 interacts with each base of RNA that is dissociated from DNA and guides RNA to the "RNA exit channel", stabilizing the elongation complex (Kent et al. 2009; Santangelo and Reeve 2010). Finally, double-stranded DNA is reformed upstream of the transcription bubble by the zipper loop (Cramer et al. 2001; Gnatt et al. 2001).

3.2.2.3. *Elongation factors*

During the elongation phase (Figure 3.5), RNAP must transcribe long genes (several thousand bp) across chromatin without dissociating from the DNA. To achieve this, RNAP must increase its processivity in order to remain stably bound to DNA. At the start of the elongation phase, the TFE factor can be retained on the clamp coiled-coil domain of RNAP, keeping the clamp domain in an open configuration, and thus decreasing the processivity of RNAP (Grohmann et al. 2011). For the elongation phase to be productive, TFE must therefore be exchanged by the elongation factor Spt4/5, which induces a conformational change in RNAP (40).

The Spt5 factor (known as NusG in bacteria) is the only transcription factor conserved in all three domains of life. The archaeal Spt5 factor contains an NGN (NusG N-terminal) domain that binds to the clamp coiled-coil domain of RNAP, as well as to its partner Spt4. Spt4 is a small protein containing a zinc finger domain that enhances the stability of Spt5 (Hirtreiter et al. 2010b). Spt5 also contains one (in bacteria and archaea) or more (in eukaryotes) C-terminal KOW (Kyrpides–Ouzounis–Woese) domains. At the start of the elongation phase, the Spt4/5 complex displaces TFE from the RNAP binding to the enzyme's clamp coiled-coil domain

(Grohmann et al. 2011). The association of Spt4/5 induces closure of the clamp domain, locking DNA into the "main channel" of RNAP (Klein et al. 2011; Martinez-Rucobo et al. 2011). Spt4/5 thus prevents dissociation between RNAP and nucleic acids, increasing TEC stability and processivity (Chakraborty et al. 2012). In bacteria, and probably in archaea, the KOW domain of Spt5/NusG interacts with the ribosomal protein S10, physically coupling the RNAP and ribosome, thus coordinating transcription and translation (Werner 2012).

Elongation Factor 1 (Elf1) is a transcriptional elongation factor conserved in eukaryotes and in several archaea species (in some TACKs (Aig-, Cren-, Kor- and Bathyarchaea) and in ASGARDs) (Daniels et al. 2009). Elf1 comprises a positively charged N-terminal region, a central zinc finger domain and an unstructured C-terminal tail (Prather et al. 2005). Elf1 surrounds the main channel of RNAP through its zinc-finger domain, and remains bound on RNAP during the elongation phase in a similar way to Spt4/5 (Mayer et al. 2010; Ehara et al. 2017). It has been shown, via ChiP-Seq, that Elf1 is recruited onto RNAP after Spt4/5 in the proximal promoter region, and that Elf1 recruitment stimulates RNAP entry into the productive elongation phase (Blombach et al. 2021). Elf1 is therefore part of the TEC and, by binding to the TEC, Elf1 acts as a barrier against physical obstacles such as DNA-bound histones, enabling the TEC to transcribe DNA across chromatin (Ehara et al. 2019).

In bacteria, the elongation factor NusA (N-utilization substance A) is a polyvalent protein that interacts with the TEC and transcribed RNA, so as to negatively or positively regulate elongation. Bacterial NusA contains the N-terminal domain that binds to RNAP, as well as a C-terminal domain that contains two KH motifs, which are ancestral nucleic acid-binding motifs. The archaeal homologue of NusA solely consists of two KH motifs that are capable of binding RNA with high affinity (Shibata et al. 2007). All archaeal species encode one or two NusA homologue genes; however, the function of archaeal NusA during elongation remains poorly understood (Fouqueau et al. 2018). The archaeal NusA factor could bind to neo-transcribed RNA and thus regulate TEC processivity by inhibiting or stimulating the formation of a secondary RNA structure.

3.2.2.4. *Arrest and reactivation of the elongation complex*

During the elongation phase, the rate of transcription is not constant. The erroneous incorporation of a non-complementary rNTP or the appearance of physical obstacles, such as histones, can frequently interrupt RNA synthesis, leading to pauses and even arrest of the TEC (Figure 3.7). Some of these "transcriptional pauses" may be brief and are resolved spontaneously by RNAP, while others may induce RNAP to retreat along the DNA (a process known as backtracking) (Nudler 2012). During backtracking, the 3′ end of the RNA dissociates from the template DNA strand, moves from the active site and exits the RNAP through the secondary channel, retaining the TEC in a state of "transcriptional arrest" (Cheung and Cramer 2011). Arrested TECs are catalytically inactive and can, in turn, act as physical barriers against other TECs or the replication machinery, compromising the maintenance of genome integrity. Rapid reactivation followed by continuation of transcription by the TEC is therefore vital for the cell. The positioning of a new 3′ end of the RNA at the active site of RNAP is essential for transcription to resume. This is achieved through the cleavage factor TFS (transcription factor S), which stimulates RNA cleavage in the RNAP active site, creating a new 3′ end of the RNA that is suitable for polymerization (Lange and Hausner 2004; Fouqueau et al. 2017b).

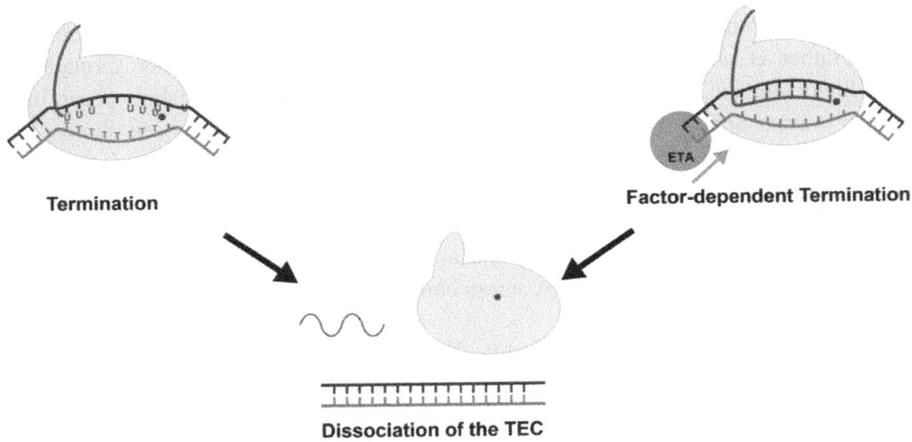

Figure 3.7. *Transcription termination and RNA release. For a color version of this figure, see www.iste.co.uk/clouet/archaea2.zip*

COMMENT ON FIGURE 3.7.– *Schematic representations of transcription termination in archaea. Termination events that do not depend on exogenous factors are called intrinsic termination. In archaea, intrinsic termination is induced by a uridine-rich RNA sequence. Unlike bacteria, secondary RNA structures do not influence the efficiency of intrinsic termination in archaea. Helicases such as ETA in archaea move along the DNA and collide with the ternary elongation complex (TEC), releasing the transcribed RNA.*

The TFS factor consists of two zinc-finger domains, bound by a flexible connector. To stimulate RNA cleavage, the C-terminal domain of TFS is positioned at the RNAP active site via the "secondary channel" (Kettenberger et al. 2004). In the RNAP active site, two negatively charged residues (aspartate and glutamate), carried by the

TFS C-terminal domain, position a second Mg^{2+} ion required for the cleavage reaction. The cleaved RNA fragment and the Mg^{2+} ion then leave the active site of RNAP. With the exception of the *Euryarchaea Methanopyrus kandleri*, all archaea species encode one or more copies of TFS (Brochier et al. 2004; Fouqueau et al. 2017b). TFS is evolutionarily linked to RPB9-like subunits and the eukaryotic transcription factor TFIIS (Langer et al. 1995). However, while the TFS and TFIIS factors combine with RNAP in a reversible manner, the RPB9-like subunits were definitively incorporated into eukaryotic RNAP during evolution (Hausner et al. 2000; Werner 2013).

3.2.3. *Termination*

Transcription termination is a key step that ensures the formation of the 3′ end of functional transcripts. This step prevents RNAP from transcribing across adjacent regions, thus avoiding transcriptional interference. Despite its biological significance, transcription termination remains one of the least understood processes of gene expression in archaea. In general terms, the mechanism of termination can be summed up to the dissociation of RNAP from the RNA molecule and DNA. Two categories of termination mechanisms have been described: sequence-dependent and factor-dependent. Sequence-dependent termination is induced by a DNA sequence called a terminator sequence. Factor-dependent termination, on the other hand, requires the intervention of external factors, often helicases or nucleases, which stimulate RNAP dissociation.

In vitro and in vivo studies have shown that *Euryarchaean* RNAP is capable of terminating transcription when it transcribes the thymidine-rich terminator sequence (poly-T) onto the non-transcribed DNA strand (Muller et al. 1985; Thomm et al. 1994). Transcription of the terminator sequence results in the formation of a particularly unstable dA-rU DNA–RNA hybrid (Sugimoto et al. 1995). The neo-synthesized RNA can then easily detach from its matrix, leading to disassociation of the TEC and to transcription termination. This intrinsic sequence termination process does not require the intervention of external factors like in the RNAP III termination process in eukaryotes (Santangelo and Reeve 2006; Spitalny and Thomm 2008; Santangelo et al. 2009; Hirtreiter et al. 2010a). Recently, corresponding sites at the 3′ ends of RNAs have been mapped in the genomes of two archaea species: *Methanosarcina mazei* (*Euryarchaea*) and *Sulfolobus acidocaldarius* (*Crenarchaea*) (Dar et al. 2016). This innovative approach, known as Term-Seq, established that transcription terminates immediately downstream of uridine-rich RNA sequences. However, the poly-T terminator sequence could only be identified respectively in 25% and 53% of the transcription units sequenced by Term-seq in *M. mazei* and *S. acidocaldarius*. This work suggests that the majority of genes and operons require either a terminator sequence other than poly-T, or the intervention of one or more external factors to complete transcription termination.

Little is known about the factors that can induce termination in archaea. In bacteria and eukaryotes, termination factors use a “torpedo” mechanism to dissociate the TEC from DNA. These factors are ribonucleases (Rat1/XRN2 in eukaryotes) or RNA helicases (e.g. Rho in bacteria; Sen1 in eukaryotes) that move along the transcribed RNA in the 5′->3′ direction and collide with the TEC, causing dissociation of the RNA and RNAP template DNA (Brennan et al. 1987; Steinmetz and Brow 1996; Kim et al. 2004; West et al. 2004). Archaeal genomes encode several proteins that could potentially induce transcription termination through the “torpedo” mechanism. The only protein that has been studied in the context of transcription termination is an RNA helicase of the Ski2 family, called ETA (euryarchaeal termination activity) or ASH (archaea-specific helicase) (Chamieh et al. 2016; Walker et al. 2017), and which is only conserved in *Euryarchaea*. The ETA helicase is recruited to the DNA, just upstream of the arrested RNAP. ETA then moves along the DNA in the 5′->3′ direction, colliding with the arrested RNAP and releasing the RNA (Walker et al. 2017). It is important to note, however, that unlike bacterial or eukaryotic termination factors, the gene encoding ETA is non-essential and only conserved in *Euryarchaea*. Furthermore, ETA is able to trigger the dissociation of arrested RNAPs, but not of active ones. Altogether, these results suggest that ETA is not a general termination factor, but rather a factor capable of dissociating RNAPs from damaged DNA regions during DNA repair.

Other proteins that could trigger the “torpedo” mechanism in archaea are ribonucleases of the β-CASP family, and in particular the ribonuclease aCPSF1 (archaeal cleavage and polyadenylation specificity factor 1). It has been demonstrated in vitro that β-CASP family member enzymes all have 5′->3′ exoribonucleolytic activity (Clouet-d’Orval et al. 2015). However, aCPSF1 additionally possesses endoribonucleolytic activity, capable of cleaving RNA at the C-A di-ribonucleotide, as well as a KH domain that enables it to bind directly to RNA (Phung et al. 2013). The aCPSF1 ribonuclease is the only member of the β-CASP family that is conserved across all archaeal species, the ortholog of the CPSF-73 subunit of the CPSF complex, a protein complex essential for cleavage and polyadenylation of messenger RNAs during transcription termination in eukaryotes (Dominski et al. 2013). Recent studies have shown that aCPSF1 (FttA) from *Thermococcus kodakarensis* is directly involved in transcription termination (Sanders et al. 2020). In vitro transcription termination experiments show the cleavage activity of aCPSF1 in cytidine-rich sequences. This activity is stimulated in the presence of the Spt4/5 factor, suggesting recruitment of aCPSF1 to RNAP via the Spt4/5 factor. Furthermore, genome-wide in vivo studies in *Methanococcus maripaludis* show that aCPSF1 depletion leads to a decrease in cell growth rate and a drastic change in RNA transcripts (Yue et al. 2020). Analysis of the 3′ end of transcripts reveals that aCPSF1 performs endoribonucleolytic cleavage downstream of a uridine-rich sequence and is essential for transcription termination. Depletion of aCPSF1 therefore leads to accumulation of read-through transcripts. Similarly, in vitro experiments have confirmed a direct interaction of aCPSF1 on uridine-rich terminator sequences through its KH domain (Li et al. 2021). This interaction results in greater efficiency of intrinsic transcription termination by these terminator sequences, regardless of the enzymatic activity of aCPSF1.

In addition to its role in transcription termination, aCPSF1 is also implicated in the premature termination of transcripts during elongation (Blombach et al. 2021). Indeed, a genome-wide analysis of the distribution of RNAPs, as well as initiation and elongation factors in *S. solfataricus* using chromatin immunoprecipitation techniques with DNA sequencing and transcriptomics, shows the cascade recruitment of elongation factors Spt4/5, Elf1 and aCPSF1 on RNAP downstream and in the vicinity of promoter regions. When the level of aCPSF1 recruitment is high, this results in a decrease in the number of initiation complexes transitioning to the elongation phase, inducing premature termination of transcripts. Intriguingly, this accumulation of aCPSF1 was observed in the majority of genes analyzed, suggesting that premature termination of transcripts is a generalized mechanism in *S. solfataricus*. This observation reveals a previously unknown mechanism of gene repression in archaea, in which RNAPs located near promoters are prevented from entering the elongation phase by a termination process regulated by a protein factor. Interestingly, in mammals, the integrator complex, whose Ins11 catalytic subunit is the direct homologue of aCPSF1, regulates cellular mRNA levels by targeting RNAP II (Elrod et al. 2019). Integrator uses its RNA endonuclease activity to destabilize RNAP II located at the promoter, thus inducing premature termination of transcripts. This observation would demonstrate that the mechanism of transcription regulation by premature termination, induced by aCPSF1 and its homologues, has been conserved throughout evolution.

Together, these recent studies describe mechanisms of transcript termination by aCPSF1, similar to mRNA termination by CPSF73 in eukaryotes. aCPSF1 does, however, appear to be capable, on its own, of stimulating intrinsic termination via terminator sequences and enzymatic activity. Furthermore, aCPSF1 is also thought to be involved in the premature termination of transcripts during transcription elongation in a similar way to the integrator complex described in eukaryotes. These multiple roles of aCPSF1 show that the protein is involved in the majority of essential transcriptional processes. Its study provides a better understanding of the evolution of transcription machinery between archaea and eukaryotes.

3.3. Transcription regulation

Transcription regulation (Figure 3.8) enables cells to respond to extra- or intracellular signals, allowing them to adapt to their environment by expressing the appropriate proteins. While in eukaryotes there are several levels of gene expression regulation (transcriptional, post-transcriptional and translational), transcription initiation regulation is the most common in archaea (Aravind and Koonin 1999; Coulson et al. 2007). This regulation is achieved through the action of transcriptional regulators that bind to specific DNA sequences, thereby affecting gene expression. The use of regulators similar to those used by bacteria could be explained by a common origin, from a single ancestor called LUCA (Last Universal Common Ancestor). However, it has been shown that the majority of archaeal transcription regulators are of bacterial origin and were obtained through horizontal gene transfer during evolution (Aravind et al. 2005; Nelson-Sathi et al. 2015).

All regulators have N-terminal DNA-binding domains and feature RHH (ribbon–helix–helix) or HTH (helix–turn–helix) motifs (Aravind et al. 2005). These regulators are classified into several families (Lrs14, MarR, TrmB, LrpA, etc.), depending on the metabolic pathway in which they are involved (Napoli et al. 1999; Brinkman et al. 2000; Lee et al. 2003; Yokoyama et al. 2006; Fiorentino et al. 2007). More often than not, they also contain a C-terminal domain that enables them to bind ligands (sugar, amino acids, etc.) and form homodimers (Schreiter and Drennan 2007; Rivera-Gomez et al. 2017). Species-specific ligand binding modulates the affinity and binding site of these regulators. Thus, these regulators play an important role in controlling and coordinating the expression of specific target genes, forming transcriptional regulatory networks and enabling the cell to respond to external stresses or metabolic requirements in an optimal manner (Martinez-Pastor et al. 2017).

Transcription regulators can be grouped into three broad functional categories: repressors, activators and dual function regulators.

In archaea, two main mechanisms of transcription repression have been described. In the first mechanism, transcriptional repressors bind the DNA region overlapping the BRE element and the TATA-box, thus blocking access to the promoter for the TFB and TBP factors by steric hindrance (Bell and Jackson 2000; Lipscomb et al. 2009). In a second mechanism, the regulator binds to a site upstream of the TATA-box, directly blocking RNAP recruitment (Figure 3.7(a)) (Bell et al. 1999a, 1999b; Dahlke and Thomm 2002). In both cases, repressors inhibit PIC formation at the target gene promoter, thus preventing gene expression.

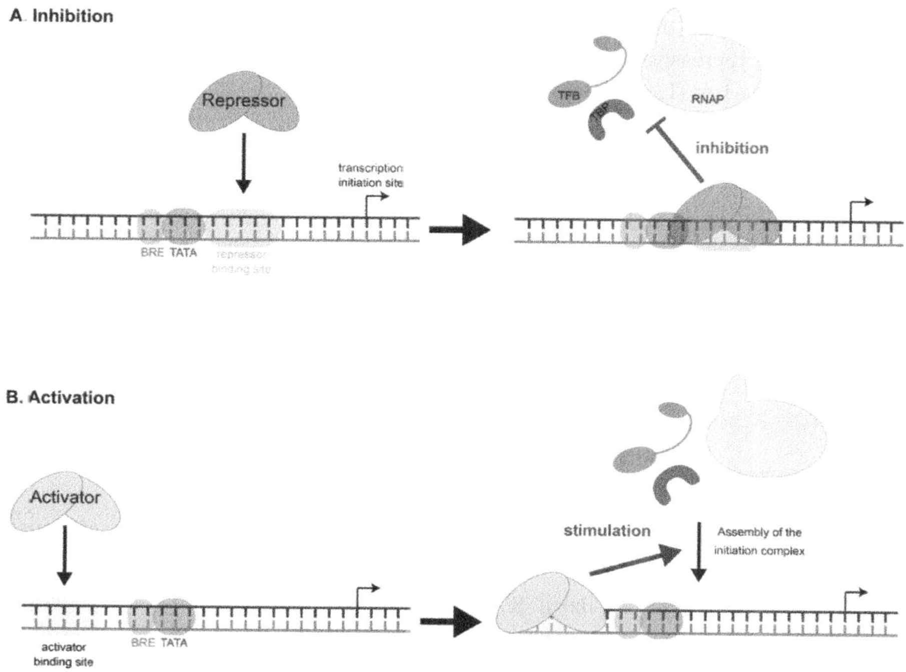

Figure 3.8. *Transcription regulation. (a) Transcription repressors bind to the promoter and physically block the recruitment of initiation factors or RNA polymerase to the promoter, thus inhibiting transcription initiation; (b) ligation of an activator upstream of a promoter stimulates the recruitment of initiation factors and RNA polymerase to the promoter, thus enabling transcription initiation. For a color version of this figure, see www.iste.co.uk/clouet/archaea2.zip*

Archaeal transcription activators stimulate the expression of target genes by promoting PIC formation at the promoter. Activators are essential for the transcription of genes with "weak" or incomplete promoters. In archaea, activators stimulate transcription by facilitating access of TBP and TFB factors to the promoter, or by stimulating recruitment of TBP or TFB factors to the "weak" promoter (Figure 3.7(b)) (Ouhammouch and Geiduschek 2005; Ouhammouch 2003; Ochs et al. 2012). These regulators bind to a DNA region upstream of the TATA-box and BRE, stimulating the recruitment of TBP or TFB factors and the stability of the TBP–TFB–DNA complex. These activators thus allow the transcription of genes, enabling PIC formation, which would not be possible without them.

It should be noted that some regulators can have dual functions, working as both activators and repressors. Indeed, it has been demonstrated in vitro that certain regulators of the Lrp (Leucine response protein) family are capable of stimulating or repressing the expression of their own gene, depending on their concentration (Peeters and Charlier 2010). Lrp can bind to three binding sites, called Box1-3, just upstream of its promoter. At low concentrations, Lrp dimers bind to Box 1 and Box 3 (high-affinity sites) and stimulate transcription of their own gene. However, at high concentrations, a Lrp dimer will also bind to Box 2, a low-affinity site. The simultaneous binding of Lrp to all three Box sites is accompanied by a cooperative protein–protein interaction that causes significant DNA deformation at the promoter level (Peeters et al. 2004, 2007). This DNA deformation affects the recruitment of TBP and TFB to the promoter, thus inhibiting PIC formation and gene expression.

3.4. References

Aravind, L. and Koonin, E.V. (1999). DNA-binding proteins and evolution of transcription regulation in the archaea. *Nucleic Acids Res.*, 27(23), 4658–4670.

Aravind, L., Anantharaman, V., Balaji, S., Babu, M.M., Iyer, L.M. (2005). The many faces of the helix–turn–helix domain: Transcription regulation and beyond. *FEMS Microbiol. Rev.*, 29(2), 231–262.

Baliga, N.S., Goo, Y.A., Ng, W.V., Hood, L., Daniels, C.J., DasSarma, S. (2000). Is gene expression in halobacterium NRC-1 regulated by multiple TBP and TFB transcription factors? *Mol. Microbiol.*, 36(5), 1184–1185.

Bar-Nahum, G., Epshtein, V., Ruckenstein, A.E., Rafikov, R., Mustaev, A., Nudler, E. (2005). A ratchet mechanism of transcription elongation and its control. *Cell*, 120(2), 183–193.

Bartlett, M.S., Thomm, M., Geiduschek, E.P. (2000). The orientation of DNA in an archaeal transcription initiation complex. *Nat. Struct. Biol.*, 7(9), 782–785.

Bell, S.D. and Jackson, S.P. (2000). Mechanism of autoregulation by an archaeal transcriptional repressor. *J. Biol. Chem.*, 275(41), 31624–31629.

Bell, S.D., Kosa, P.L., Sigler, P.B., Jackson, S.P. (1999a). Orientation of the transcription preinitiation complex in archaea. *Proc. Natl. Acad. Sci. USA*, 96(24), 13662–13667.

Bell, S.D., Cairns, S.S., Robson, R.L., Jackson, S.P. (1999b). Transcriptional regulation of an archaeal operon in vivo and in vitro. *Mol. Cell.*, 4(6), 971–982.

Bell, S.D., Magill, C.P., Jackson, S.P. (2001a). Basal and regulated transcription in Archaea. *Biochem. Soc. Trans.*, 29(4), 392–395.

Bell, S.D., Brinkman, A.B., van der Oost, J., Jackson, S.P. (2001b). The archaeal TFIIEalpha homologue facilitates transcription initiation by enhancing TATA-box recognition. *EMBO Rep.*, 2(2), 133–138.

Bewley, C.A., Gronenborn, A.M., Clore, G.M. (1998). Minor groove-binding architectural proteins: Structure, function, and DNA recognition. *Annu. Rev. Biophys. Biomol. Struct.*, 27, 105–131.

Bischof, L.F., Haurat, M.F., Hoffmann, L., Albersmeier, A., Wolf, J., Neu, A., Pham, T.K., Albaum, S.P., Jakobi, T., Schouten, S. et al. (2018). Early response of *Sulfolobus acidocaldarius* to nutrient limitation. *Front Microbiol.*, 9, 3201.

Blombach, F., Salvadori, E., Fouqueau, T., Yan, J., Reimann, J., Sheppard, C., Smollett, K.L., Albers, S.V., Kay, C.W.M., Thalassinos, K. et al. (2015). Archaeal TFEalpha/beta is a hybrid of TFIIE and the RNA polymerase III subcomplex hRPC62/39. *Elife*, 4, e08378.

Blombach, F., Smollett, K.L., Grohmann, D., Werner, F. (2016). Molecular mechanisms of transcription initiation-structure, function, and evolution of TFE/TFIIE-like factors and open complex formation. *J. Mol. Biol*, 428(12), 2592–2606.

Blombach, F., Ausiannikava, D., Figueiredo, A.M., Soloviev, Z., Prentice, T., Zhang, M., Zhou, N., Thalassinos, K., Allers, T., Werner, F. (2018). Structural and functional adaptation of *Haloferax volcanii* TFEalpha/beta. *Nucleic Acids Res.*, 46(5), 2308–2320.

Blombach, F., Fouqueau, T., Matelska, D., Smollett, K., Werner, F. (2021). Promoter–proximal elongation regulates transcription in archaea. *Nat. Commun.*, 12(1), 5524.

Brennan, C.A., Dombroski, A.J., Platt, T. (1987). Transcription termination factor rho is an RNA–DNA helicase. *Cell*, 48(6), 945–952.

Brinkman, A.B., Dahlke, I., Tuininga, J.E., Lammers, T., Dumay, V., de Heus, E., Lebbink, J.H., Thomm, M., de Vos, W.M., van Der Oost, J. (2000). An Lrp-like transcriptional regulator from the archaeon *Pyrococcus furiosus* is negatively autoregulated. *J. Biol. Chem.*, 275(49), 38160–38169.

Brochier, C., Forterre, P., Gribaldo, S. (2004). Archaeal phylogeny based on proteins of the transcription and translation machineries: Tackling the *Methanopyrus kandleri* paradox. *Genome Biol.*, 5(3), R17.

Brueckner, F. and Cramer, P. (2008). Structural basis of transcription inhibition by alpha-amanitin and implications for RNA polymerase II translocation. *Nat. Struct. Mol. Biol.*, 15(8), 811–818.

Chakraborty, A., Wang, D., Ebright, Y.W., Korlann, Y., Kortkhonjia, E., Kim, T., Chowdhury, S., Wigneshweraraj, S., Irschik, H., Jansen, R. et al. (2012). Opening and closing of the bacterial RNA polymerase clamp. *Science*, 337(6094), 591–595.

Chamieh, H., Ibrahim, H., Kozah, J. (2016). Genome-wide identification of SF1 and SF2 helicases from archaea. *Gene*, 576(1/2), 214–228.

Cheung, A.C. and Cramer, P. (2011). Structural basis of RNA polymerase II backtracking, arrest and reactivation. *Nature*, 471(7337), 249–253.

Clouet-d'Orval, B., Phung, D.K., Langendijk-Genevaux, P.S., Quentin, Y. (2015). Universal RNA-degrading enzymes in archaea: Prevalence, activities and functions of beta-CASP ribonucleases. *Biochimie*, 118, 278–285.

Coulson, R.M., Touboul, N., Ouzounis, C.A. (2007). Lineage-specific partitions in archaeal transcription. *Archaea*, 2(2), 117–125.

Cramer, P. (2002). Multisubunit RNA polymerases. *Curr. Opin. Struct. Biol.*, 12(1), 89–97.

Cramer, P., Bushnell, D.A., Kornberg, R.D. (2001). Structural basis of transcription: RNA polymerase II at 2.8 angstrom resolution. *Science*, 292(5523), 1863–1876.

Dahlke, I. and Thomm, M. (2002). A Pyrococcus homolog of the leucine-responsive regulatory protein, LrpA, inhibits transcription by abrogating RNA polymerase recruitment. *Nucleic Acids Res.*, 30(3), 701–710.

Daniels, J.P., Kelly, S., Wickstead, B., Gull, K. (2009). Identification of a crenarchaeal orthologue of Elf1: Implications for chromatin and transcription in *Archaea. Biol. Direct.*, 4(24). doi: 10.1186/1745-6150-4-24.

Dar, D., Prasse, D., Schmitz, R.A., Sorek, R. (2016). Widespread formation of alternative 3' UTR isoforms via transcription termination in archaea. *Nat. Microbiol.*, 1(10), 16143.

Dexl, S., Reichelt, R., Kraatz, K., Schulz, S., Grohmann, D., Bartlett, M., Thomm, M. (2018). Displacement of the transcription factor B reader domain during transcription initiation. *Nucleic Acids Res.*, 46(19), 10066–10081. doi: 10.1093/nar/gky699.

Dominski, Z., Carpousis, A.J., Clouet-d'Orval, B. (2013). Emergence of the beta-CASP ribonucleases: Highly conserved and ubiquitous metallo-enzymes involved in messenger RNA maturation and degradation. *Biochim. Biophys. Acta.*, 1829(6/7), 532–551.

Ehara, H., Yokoyama, T., Shigematsu, H., Yokoyama, S., Shirouzu, M., Sekine, S.I. (2017). Structure of the complete elongation complex of RNA polymerase II with basal factors. *Science*, 357(6354), 921–924.

Ehara, H., Kujirai, T., Fujino, Y., Shirouzu, M., Kurumizaka, H., Sekine, S.I. (2019). Structural insight into nucleosome transcription by RNA polymerase II with elongation factors. *Science*, 363(6428), 744–747.

Elrod, N.D., Henriques, T., Huang, K.L., Tatomer, D.C., Wilusz, J.E., Wagner, E.J., Adelman, K. (2019). The integrator complex attenuates promoter-proximal transcription at protein-coding genes. *Mol. Cell.*, 76(5), 738–752.e7.

Facciotti, M.T., Reiss, D.J., Pan, M., Kaur, A., Vuthoori, M., Bonneau, R., Shannon, P., Srivastava, A., Donohoe, S.M., Hood, L.E. et al. (2007). General transcription factor specified global gene regulation in archaea. *Proc. Natl. Acad. Sci. USA*, 104(11), 4630–4635.

Fiorentino, G., Ronca, R., Cannio, R., Rossi, M., Bartolucci, S. (2007). MarR-like transcriptional regulator involved in detoxification of aromatic compounds in *Sulfolobus solfataricus*. *J. Bacteriol.*, 189(20), 7351–7360.

Fouqueau, T., Zeller, M.E., Cheung, A.C., Cramer, P., Thomm, M. (2013). The RNA polymerase trigger loop functions in all three phases of the transcription cycle. *Nucleic Acids Res.*, 41(14), 7048–7059.

Fouqueau, T., Blombach, F., Werner, F. (2017a). Evolutionary origins of two-barrel RNA polymerases and site-specific transcription initiation. *Annu. Rev. Microbiol.*, 71, 331–348.

Fouqueau, T., Blombach, F., Hartman, R., Cheung, A.C.M., Young, M.J., Werner, F. (2017b). The transcript cleavage factor paralogue TFS4 is a potent RNA polymerase inhibitor. *Nat. Commun.*, 8(1), 1914.

Fouqueau, T., Blombach, F., Cackett, G., Carty, A.E., Matelska, D.M., Ofer, S., Pilotto, S., Phung, D.K., Werner, F. (2018). The cutting edge of archaeal transcription. *Emerging Topics in Life Sciences*, 2(4), 517.

Gietl, A., Holzmeister, P., Blombach, F., Schulz, S., von Voithenberg, L.V., Lamb, D.C., Werner, F., Tinnefeld, P., Grohmann, D. (2014). Eukaryotic and archaeal TBP and TFB/TF(II)B follow different promoter DNA bending pathways. *Nucleic Acids Res.*, 42(10), 6219–6231.

Gnatt, A.L., Cramer, P., Fu, J., Bushnell, D.A., Kornberg, R.D. (2001). Structural basis of transcription: An RNA polymerase II elongation complex at 3.3 A resolution. *Science*, 292(5523), 1876–1882.

Grohmann, D. and Werner, F. (2011). Cycling through transcription with the RNA polymerase F/E (RPB4/7) complex: Structure, function and evolution of archaeal RNA polymerase. *Res. Microbiol.*, 162(1), 10–18.

Grohmann, D., Nagy, J., Chakraborty, A., Klose, D., Fielden, D., Ebright, R.H., Michaelis, J., Werner, F. (2011). The initiation factor TFE and the elongation factor Spt4/5 compete for the RNAP clamp during transcription initiation and elongation. *Mol. Cell.*, 43(2), 263–274.

Hanzelka, B.L., Darcy, T.J., Reeve, J.N. (2001). TFE, an archaeal transcription factor in *Methanobacterium thermocutotrophicum* related to eucaryal transcription factor TFIIEalpha. *J. Bacteriol.*, 183(5), 1813–1818.

Hausner, W., Wettach, J., Hethke, C., Thomm, M. (1996). Two transcription factors related with the eucaryal transcription factors TATA-binding protein and transcription factor IIB direct promoter recognition by an archaeal RNA polymerase. *J. Biol. Chem.*, 271(47), 30144–30148.

Hausner, W., Lange, U., Musfeldt, M. (2000). Transcription factor S, a cleavage induction factor of the archaeal RNA polymerase. *J. Biol. Chem.*, 275(17), 12393–12399.

Hirtreiter, A., Grohmann, D., Werner, F. (2010a). Molecular mechanisms of RNA polymerase – The F/E (RPB4/7) complex is required for high processivity in vitro. *Nucleic Acids Res.*, 38(2), 585–596.

Hirtreiter, A., Damsma, G.E., Cheung, A.C., Klose, D., Grohmann, D., Vojnic, E., Martin, A.C.R., Cramer, P., Werner, F. (2010b). Spt4/5 stimulates transcription elongation through the RNA polymerase clamp coiled-coil motif. *Nucleic Acids Res.*, 38(12), 4040–4051.

Iyer, L.M., Koonin, E.V., Aravind, L. (2003). Evolutionary connection between the catalytic subunits of DNA-dependent RNA polymerases and eukaryotic RNA-dependent RNA polymerases and the origin of RNA polymerases. *BMC Struct. Biol.*, 3(1). doi: 10.1186/1472-6807-3-1.

Kanai, T., Akerboom, J., Takedomi, S., van de Werken, H.J., Blombach, F., van der Oost, J., Murakami, T., Atomi, H., Imanaka, T. (2007). A global transcriptional regulator in *Thermococcus kodakaraensis* controls the expression levels of both glycolytic and gluconeogenic enzyme-encoding genes. *J. Biol. Chem.*, 282(46), 33659–33670.

Kaplan, C.D., Larsson, K.M., Kornberg, R.D. (2008). The RNA polymerase II trigger loop functions in substrate selection and is directly targeted by alpha-amanitin. *Mol. Cell.*, 30(5), 547–556.

Kent, T., Kashkina, E., Anikin, M., Temiakov, D. (2009). Maintenance of RNA–DNA hybrid length in bacterial RNA polymerases. *J. Biol. Chem.*, 284(20), 13497–13504.

Kettenberger, H., Armache, K.J., Cramer, P. (2004). Complete RNA polymerase II elongation complex structure and its interactions with NTP and TFIIS. *Mol. Cell.*, 16(6), 955–965.

Kim, M., Krogan, N.J., Vasiljeva, L., Rando, O.J., Nedea, E., Greenblatt, J.F., Buratowski, S. (2004). The yeast Rat1 exonuclease promotes transcription termination by RNA polymerase II. *Nature*, 432(7016), 517–522.

Kireeva, M.L., Komissarova, N., Waugh, D.S., Kashlev, M. (2000). The 8-nucleotide-long RNA:DNA hybrid is a primary stability determinant of the RNA polymerase II elongation complex. *J. Biol. Chem.*, 275(9), 6530–6536.

Klein, B.J., Bose, D., Baker, K.J., Yusoff, Z.M., Zhang, X., Murakami, K.S. (2011). RNA polymerase and transcription elongation factor Spt4/5 complex structure. *Proc. Natl. Acad. Sci. USA*, 108(2), 546–550.

Korkhin, Y., Unligil, U.M., Littlefield, O., Nelson, P.J., Stuart, D.I., Sigler, P.B., Bell, S.D., Abrescia, N.G.A. (2009). Evolution of complex RNA polymerases: The complete archaeal RNA polymerase structure. *PLoS Biol.*, 7(5), e1000102.

Kosa, P.F., Ghosh, G., DeDecker, B.S., Sigler, P.B. (1997). The 2.1-A crystal structure of an archaeal preinitiation complex: TATA-box-binding protein/transcription factor (II)B core/TATA-box. *Proc. Natl. Acad. Sci. USA*, 94(12), 6042–6047.

Kostrewa, D., Zeller, M.E., Armache, K.J., Seizl, M., Leike, K., Thomm, M., Cramer, P. (2009). RNA polymerase II-TFIIB structure and mechanism of transcription initiation. *Nature*, 462(7271), 323–330.

Kuznedelov, K., Korzheva, N., Mustaev, A., Severinov, K. (2002). Structure-based analysis of RNA polymerase function: The largest subunit's rudder contributes critically to elongation complex stability and is not involved in the maintenance of RNA–DNA hybrid length. *EMBO J.*, 21(6), 1369–1378.

Lane, W.J. and Darst, S.A. (2010a). Molecular evolution of multisubunit RNA polymerases: Sequence analysis. *J. Mol. Biol.*, 395(4), 671–685.

Lane, W.J. and Darst, S.A. (2010b). Molecular evolution of multisubunit RNA polymerases: Structural analysis. *J. Mol. Biol.*, 395(4), 686–704.

Lange, U. and Hausner, W. (2004). Transcriptional fidelity and proofreading in Archaea and implications for the mechanism of TFS-induced RNA cleavage. *Mol. Microbiol.*, 52(4), 1133–1143.

Langer, D., Hain, J., Thuriaux, P., Zillig, W. (1995). Transcription in Archaea – Similarity to that in Eucarya. *Proc. Natl. Acad. Sci. USA*, 92(13), 5768–5772.

Lee, S.J., Engelmann, A., Horlacher, R., Qu, Q., Vierke, G., Hebbeln, C., Thomm, M., Boos, W. (2003). TrmB, a sugar-specific transcriptional regulator of the trehalose/maltose ABC transporter from the hyperthermophilic archaeon *Thermococcus litoralis*. *J. Biol Chem.*, 278(2), 983–990.

Li, J., Yue, L., Li, Z., Zhang, W., Zhang, B., Zhao, F., Dong, X. (2021). aCPSF1 cooperates with terminator U-tract to dictate archaeal transcription termination efficacy. *Elife*, 10, e70464. doi: 10.7554/eLife.70464.

Lipscomb, G.L., Keese, A.M., Cowart, D.M., Schut, G.J., Thomm, M., Adams, M.W., Scott, R.A. (2009). SurR: A transcriptional activator and repressor controlling hydrogen and elemental sulphur metabolism in *Pyrococcus furiosus*. *Mol. Microbiol.*, 71(2), 332–349.

Markovtsov, V., Mustaev, A., Goldfarb, A. (1996). Protein–RNA interactions in the active center of transcription elongation complex. *Proc. Natl. Acad. Sci. USA*, 93(8), 3221–3226.

Martinez-Pastor, M., Tonner, P.D., Darnell, C.L., Schmid, A.K. (2017). Transcriptional regulation in Archaea: From individual genes to global regulatory networks. *Annu. Rev. Genet.*, 51, 143–170.

Martinez-Rucobo, F.W., Sainsbury, S., Cheung, A.C., Cramer, P. (2011). Architecture of the RNA polymerase-Spt4/5 complex and basis of universal transcription processivity. *EMBO J.*, 30(7), 1302–1310.

Mayer, A., Lidschreiber, M., Siebert, M., Leike, K., Soding, J., Cramer, P. (2010). Uniform transitions of the general RNA polymerase II transcription complex. *Nat. Struct. Mol. Biol.*, 17(10), 1272–1278.

Meinhart, A., Blobel, J., Cramer, P. (2003). An extended winged helix domain in general transcription factor E/IIE alpha. *J. Biol. Chem.*, 278(48), 48267–48274.

Muller, B., Allmansberger, R., Klein, A. (1985). Termination of a transcription unit comprising highly expressed genes in the archaebacterium *Methanococcus voltae*. *Nucleic Acids Res.*, 13(18), 6439–6445.

Naji, S., Grünberg, S., Thomm, M. (2007). The RPB7 orthologue E' is required for transcriptional activity of a reconstituted archaeal core enzyme at low temperatures and stimulates open complex formation. *J. Biol. Chem.*, 282(15), 11047–11057.

Naji, S., Bertero, M.G., Spitalny, P., Cramer, P., Thomm, M. (2008). Structure-function analysis of the RNA polymerase cleft loops elucidates initial transcription, DNA unwinding and RNA displacement. *Nucleic Acids Res.*, 36(2), 676–687.

Napoli, A., van der Oost, J., Sensen, C.W., Charlebois, R.L., Rossi, M., Ciaramella, M. (1999). An Lrp-like protein of the hyperthermophilic archaeon *Sulfolobus solfataricus* which binds to its own promoter. *J. Bacteriol.*, 181(5),1474–1480.

Naryshkina, T., Kuznedelov, K., Severinov, K. (2006). The role of the largest RNA polymerase subunit lid element in preventing the formation of extended RNA–DNA hybrid. *J. Mol. Biol.*, 361(4), 634–643.

Nelson-Sathi, S., Sousa, F.L., Roettger, M., Lozada-Chavez, N., Thiergart, T., Janssen, A., Bryant, D., Landan, G., Schönheit, P., Siebers, B. et al. (2015). Origins of major archaeal clades correspond to gene acquisitions from bacteria. *Nature*, 517(7532), 77–80.

Nikolov, D.B., Chen, H., Halay, E.D., Usheva, A.A., Hisatake, K., Lee, D.K., Roeder, R.G., Burley, S.K. (1995). Crystal structure of a TFIIB–TBP–TATA–element ternary complex. *Nature*, 377(6545), 119–128.

Nudler, E. (1999). Transcription elongation: Structural basis and mechanisms. *J. Mol. Biol.*, 288(1), 1–12.

Nudler, E. (2012). RNA polymerase backtracking in gene regulation and genome instability. *Cell*, 149(7), 1438–1445.

Ochs, S.M., Thumann, S., Richau, R., Weirauch, M.T., Lowe, T.M., Thomm, M., Hausner, W. (2012). Activation of archaeal transcription mediated by recruitment of transcription factor B. *J. Biol. Chem.*, 287(22), 18863–18871.

Ouhammouch, M. and Geiduschek, E.P. (2005). An expanding family of archaeal transcriptional activators. *Proc. Natl. Acad. Sci. USA*, 102(43), 15423–15428.

Ouhammouch, M., Dewhurst, R.E., Hausner, W., Thomm, M., Geiduschek, E.P. (2003). Activation of archaeal transcription by recruitment of the TATA-binding protein. *Proc. Natl. Acad. Sci. USA*, 100(9), 5097–5102.

Patikoglou, G.A., Kim, J.L., Sun, L., Yang, S.H., Kodadek, T., Burley, S.K. (1999). TATA element recognition by the TATA box-binding protein has been conserved throughout evolution. *Genes Dev.*, 13(24), 3217–3230.

Peeters, E. and Charlier, D. (2010). The Lrp family of transcription regulators in archaea. *Archaea*, 2010(1), 750457. doi: 10.1155/2010/750457.

Peeters, E., Thia-Toong, T.L., Gigot, D., Maes, D., Charlier, D. (2004). Ss-LrpB, a novel Lrp-like regulator of *Sulfolobus solfataricus* P2, binds cooperatively to three conserved targets in its own control region. *Mol. Microbiol.*, 54(2), 321–336.

Peeters, E., Wartel, C., Maes, D., Charlier, D. (2007). Analysis of the DNA-binding sequence specificity of the archaeal transcriptional regulator Ss-LrpB from *Sulfolobus solfataricus* by systematic mutagenesis and high resolution contact probing. *Nucleic Acids Res.*, 35(2), 623–633.

Phung, D.K., Rinaldi, D., Langendijk-Genevaux, P.S., Quentin, Y., Carpousis, A.J., Clouet-d'Orval, B. (2013). Archaeal beta-CASP ribonucleases of the aCPSF1 family are orthologs of the eukaryal CPSF-73 factor. *Nucleic Acids Res.*, 41(2), 1091–1103.

Prather, D., Krogan, N.J., Emili, A., Greenblatt, J.F., Winston, F. (2005). Identification and characterization of Elf1, a conserved transcription elongation factor in *Saccharomyces cerevisiae. Mol. Cell. Biol.*, 25(22), 10122–10135.

Reiter, W.D., Palm, P., Zillig, W. (1988). Analysis of transcription in the archaebacterium Sulfolobus indicates that archaebacterial promoters are homologous to eukaryotic pol II promoters. *Nucleic Acids Res.*, 16(1), 1–19.

Revyakin, A., Liu, C., Ebright, R.H., Strick, T.R. (2006). Abortive initiation and productive initiation by RNA polymerase involve DNA scrunching. *Science*, 314(5802), 1139–1143.

Rivera-Gomez, N., Martinez-Nunez, M.A., Pastor, N., Rodriguez-Vazquez, K., Perez-Rueda, E. (2017). Dissecting the protein architecture of DNA-binding transcription factors in bacteria and archaea. *Microbiology*, 163(8), 1167–1178.

Rowlands, T., Baumann, P., Jackson, S.P. (1994). The TATA-binding protein: A general transcription factor in eukaryotes and archaebacteria. *Science*, 264(5163), 1326–1329.

Sainsbury, S., Niesser, J., Cramer, P. (2013). Structure and function of the initially transcribing RNA polymerase II-TFIIB complex. *Nature*, 493(7432), 437–440.

Salgado, P.S., Koivunen, M.R., Makeyev, E.V., Bamford, D.H., Stuart, D.I., Grimes, J.M. (2006). The structure of an RNAi polymerase links RNA silencing and transcription. *PLoS Biol.*, 4(12), e434.

Sanders, T.J., Wenck, B.R., Selan, J.N., Barker, M.P., Trimmer, S.A., Walker, J.E., Santangelo, T.J. (2020). FttA is a CPSF73 homologue that terminates transcription in Archaea. *Nat. Microbiol.*, 5(4), 545–553.

Santangelo, T.J. and Reeve, J.N. (2006). Archaeal RNA polymerase is sensitive to intrinsic termination directed by transcribed and remote sequences. *J. Mol. Biol.*, 355(2), 196–210.

Santangelo, T.J. and Reeve, J.N. (2010). Deletion of switch 3 results in an archaeal RNA polymerase that is defective in transcript elongation. *J. Biol. Chem.*, 285(31), 23908–23915.

Santangelo, T.J., Cubonova, L., Skinner, K.M., Reeve, J.N. (2009). Archaeal intrinsic transcription termination in vivo. *J. Bacteriol.*, 191(22), 7102–7108.

Sauguet, L. (2019). The extended "two-barrel" polymerases superfamily: Structure, function and evolution. *J. Mol. Biol.*, 431(20), 4167–4183. doi: 10.1016/j.jmb.2019.05.017.

Sauguet, L., Raia, P., Henneke, G., Delarue, M. (2016). Shared active site architecture between archaeal PolD and multi-subunit RNA polymerases revealed by X-ray crystallography. *Nat. Commun.*, 7, 12227.

Schreiter, E.R. and Drennan, C.L. (2007). Ribbon–helix–helix transcription factors: Variations on a theme. *Nat. Rev. Microbiol.*, 5(9), 710–720.

Schulz, S., Gietl, A., Smollett, K., Tinnefeld, P., Werner, F., Grohmann, D. (2016). TFE and Spt4/5 open and close the RNA polymerase clamp during the transcription cycle. *Proc. Natl. Acad. Sci. USA*, 113(13), E1816–E825.

Shibata, R., Bessho, Y., Shinkai, A., Nishimoto, M., Fusatomi, E., Terada, T., Shirouzu, M., Yokoyama, S. (2007). Crystal structure and RNA-binding analysis of the archaeal transcription factor NusA. *Biochem. Biophys. Res. Commun.*, 355(1), 122–128.

Smollett, K., Blombach, F., Reichelt, R., Thomm, M., Werner, F. (2017). A global analysis of transcription reveals two modes of Spt4/5 recruitment to archaeal RNA polymerase. *Nat. Microbiol.*, 2, 17021.

Spitalny, P. and Thomm, M. (2008). A polymerase III-like reinitiation mechanism is operating in regulation of histone expression in archaea. *Mol. Microbiol.*, 67(5), 958–970.

Steinmetz, E.J. and Brow, D.A. (1996). Repression of gene expression by an exogenous sequence element acting in concert with a heterogeneous nuclear ribonucleoprotein-like protein, Nrd1, and the putative helicase Sen1. *Mol. Cell. Biol.*, 16(12), 6993–7003.

Steitz, T.A. (1998). A mechanism for all polymerases. *Nature*, 391(6664), 231–232.

Sugimoto, N., Nakano, S., Katoh, M., Matsumura, A., Nakamuta, H., Ohmichi, T., Yoneyama, M., Sasaki, M. (1995). Thermodynamic parameters to predict stability of RNA/DNA hybrid duplexes. *Biochemistry*, 34(35), 11211–11216.

Tan, L., Wiesler, S., Trzaska, D., Carney, H.C., Weinzierl, R.O. (2008). Bridge helix and trigger loop perturbations generate superactive RNA polymerases. *J. Biol.*, 7(10), 40.

Thomm, M. and Wich, G. (1988). An archaebacterial promoter element for stable RNA genes with homology to the TATA box of higher eukaryotes. *Nucleic Acids. Res.*, 16(1), 151–163.

Thomm, M., Hausner, W., Hethke, C. (1994). Transcription factors and termination of transcription in *Methanococcus*. *Syst. Appl. Microbiol.*, 16(4), 648–655.

Toulokhonov, I. and Landick, R. (2006). The role of the lid element in transcription by *E. coli* RNA polymerase. *J. Mol. Biol.*, 361(4), 644–658.

Tuske, S., Sarafianos, S.G., Wang, X., Hudson, B., Sineva, E., Mukhopadhyay, J., Birktoft, J.J., Leroy, O., Ismail, S., Clark Jr, A.D. et al. (2005). Inhibition of bacterial RNA polymerase by streptolydigin: Stabilization of a straight–bridge–helix active-center conformation. *Cell*, 122(4), 541–552.

Vassylyev, D.G., Vassylyeva, M.N., Perederina, A., Tahirov, T.H., Artsimovitch, I. (2007a). Structural basis for transcription elongation by bacterial RNA polymerase. *Nature*, 448(7150), 157–162.

Vassylyev, D.G., Vassylyeva, M.N., Zhang, J., Palangat, M., Artsimovitch, I., Landick, R. (2007b). Structural basis for substrate loading in bacterial RNA polymerase. *Nature*, 448(7150), 163–168.

Walker, J.E., Luyties, O., Santangelo, T.J. (2017). Factor-dependent archaeal transcription termination. *Proc. Natl. Acad. Sci. USA*, 114(33), E6767–E6773.

Wang, D., Bushnell, D.A., Westover, K.D., Kaplan, C.D., Kornberg, R.D. (2006). Structural basis of transcription: Role of the trigger loop in substrate specificity and catalysis. *Cell*, 127(5), 941–954.

Weinzierl, R.O. (2010). The nucleotide addition cycle of RNA polymerase is controlled by two molecular hinges in the Bridge Helix domain. *BMC Biol.*, 8, 134. doi: 10.1186/1741-7007-8-134.

Werner, F. (2012). A nexus for gene expression-molecular mechanisms of Spt5 and NusG in the three domains of life. *J. Mol. Biol.*, 417(1/2), 13–27.

Werner, F. (2013). Molecular mechanisms of transcription elongation in archaea. *Chem. Rev.*, 113(11), 8331–8349.

Werner, F. and Grohmann, D. (2011). Evolution of multisubunit RNA polymerases in the three domains of life. *Nat. Rev. Microbiol.*, 9(2), 85–98.

Werner, F. and Weinzierl, R.O. (2002). A recombinant RNA polymerase II-like enzyme capable of promoter-specific transcription. *Mol. Cell.*, 10(3), 635–646.

Werner, F. and Weinzierl, R.O (2005). Direct modulation of RNA polymerase core functions by basal transcription factors. *Mol. Cell. Biol.*, 25(18), 8344–8355.

West, S., Gromak, N., Proudfoot, N.J. (2004). Human 5' –> 3' exonuclease Xrn2 promotes transcription termination at co-transcriptional cleavage sites. *Nature*, 432(7016), 522–525.

Whittington, J.E., Delgadillo, R.F., Attebury, T.J., Parkhurst, L.K., Daugherty, M.A., Parkhurst, L.J. (2008). TATA-binding protein recognition and bending of a consensus promoter are protein species dependent. *Biochemistry*, 47(27), 7264–7273.

Yokoyama, K., Ishijima, S.A., Clowney, L., Koike, H., Aramaki, H., Tanaka, C., Makino, K., Suzuki, M. (2006). Feast/famine regulatory proteins (FFRPs): *Escherichia coli* Lrp, AsnC and related archaeal transcription factors. *FEMS Microbiol. Rev.*, 30(1), 89–108.

Yue, L., Li, J., Zhang, B., Qi, L., Li, Z., Zhao, F., Li, L., Zheng, X., Dong, X. (2020). The conserved ribonuclease aCPSF1 triggers genome-wide transcription termination of Archaea via a 3'-end cleavage mode. *Nucleic Acids Res.*, 48(17), 9589–9605.

Yuzenkova, Y., Bochkareva, A., Tadigotla, V.R., Roghanian, M., Zorov, S., Severinov, K., Zenkin, N. (2010). Stepwise mechanism for transcription fidelity. *BMC Biol.*, 8, 54. doi: 10.1186/1741-7007-8-54.

Zaychikov, E., Martin, E., Denissova, L., Kozlov, M., Markovtsov, V., Kashlev, M., Heumann, H., Nikiforov, V., Goldfarb, A., Mustaev, A. (1996). Mapping of catalytic residues in the RNA polymerase active center. *Science*, 273(5271), 107–109.

4

RNA Classes and Their Maturation and Degradation Enzymes

Manon BATISTA, Béatrice CLOUET-D'ORVAL and Marie BOUVIER
MCD-CBI, CNRS, Université de Toulouse, France

4.1. Introduction

Cells adapt their growth and metabolism to a changing environment by regulating the expression of their genes. As RNA is the only direct product of the genome, its stability and integrity condition gene expression. Defects in the degradation and quality control of these RNAs affect cellular homeostasis, by allowing the appearance of aberrant, defective or overabundant RNA species (Wolin and Maquat 2019). Post-transcriptional RNA maturation and degradation pathways regulate the stability of messenger RNAs and regulatory noncoding RNAs, giving rise to the mature forms of ribosomal RNAs and transfer RNAs. These pathways involve molecular machines, with ribonucleases (RNases) at their core that attack the phosphodiester bonds of RNA molecules at their 5′ or 3′ ends (exoribonucleases), or within the polynucleotide chain (endoribonucleases). The ability of RNases to degrade a RNA depends on both its sequence and structure, as well as on regulatory molecules such as RNA helicases and noncoding RNAs. The multiple RNA degradation and surveillance pathways discovered in bacteria and eukaryotes highlight the complexity of these processes (Stoecklin and Mühlemann 2013). Studying the differences and similarities between the RNA degradation machines of bacteria and eukaryotes has shed light on how they work.

Biology of Archaea 2,
coordinated by Béatrice CLOUET-D'ORVAL, Bruno FRANZETTI and Philippe OGER.

In archaea, research on RNA processing and degradation pathways is still in its early phases (Clouet-d'Orval et al. 2018). In the absence of easy genetic approaches, the study of these pathways is often based on genomic and proteomic analyses that are interpreted from our understanding of RNA metabolism in bacteria and eukaryotes. The majority of ribonucleases have been researched on the basis of their homologies with those of bacteria and eukaryotes. To date, no archaea-specific ribonucleases have been identified. Double-stranded RNA-cleaving endoribonuclease activity is also still missing. From these studies, it appears that the specificities of archaeal RNAs, and the processes that finely regulate their qualities and stabilities, possess mosaic eukaryotic and bacterial characteristics (Figure 4.1).

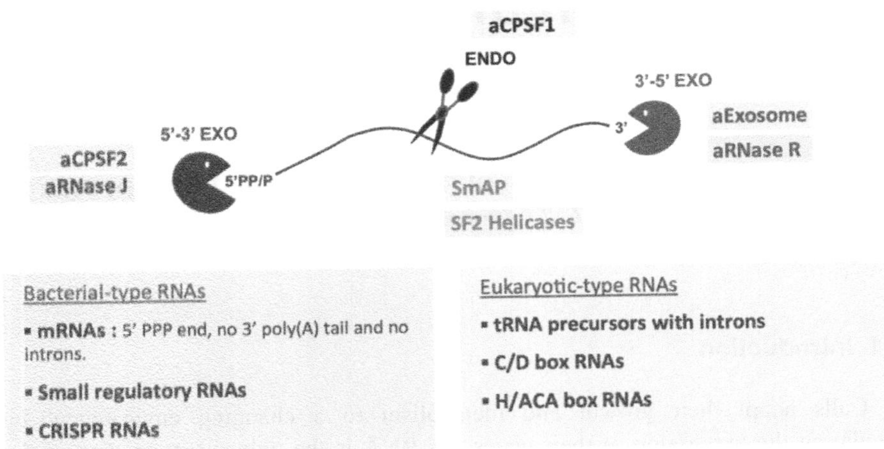

Figure 4.1. *Metabolism in archaea: a mosaic of bacterial and eukaryotic characteristics. For a color version of this figure, see www.iste.co.uk/clouet/archaea2.zip*

COMMENT ON FIGURE 4.1. – *Simplified diagram of RNA degradation representing some of the players in RNA metabolism in archaea. Bacterial and eukaryotic players are highlighted in yellow and light green, respectively. The 5'–3' and 3'–5' exoribonucleases are shown in blue and green, respectively. The endoribonuclease aCPSF1 (which also has 5'–3' exo activity) and RNA-binding proteins are shown in red and purple, respectively.*

The aim of this chapter is to give an overview of our knowledge of RNA metabolism in archaea, with a particular focus on RNA maturation and degradation enzymes. It is important to note that observations made in some groups of archaea cannot be generalized to all archaea.

4.2. RNA classes in archaea

RNA transcription is handled by a single RNA polymerase related to eukaryotic RNA polymerase II (see Chapter 3 of this volume). As in all living cells, the expression of genetic information involves three main types of RNA: messenger RNAs (mRNAs), which define the proteome; stable RNAs, which group together ribosomal RNAs (rRNAs) and transfer RNAs (tRNAs), functioning together to decode mRNA sequences; and lastly noncoding regulatory RNAs (ncRNAs), which regulate the stability and/or the ability of an mRNA to be translated, and guide the addition of chemical modifications (box C/D and H/ACA guide RNA panel). Within these classes, archaeal RNAs possess characteristics that are typical of eukaryotes and bacteria.

4.2.1. *Messenger RNA*

Archaeal mRNAs are of bacterial type (Figure 4.2): they have no introns or a 3′ poly(A) tail; their triphosphorylated 5′ ends are unmodified and can, in some cases, have a riboswitch structure (Gupta and Swati 2019); they can be polycistronic; the 5′-UTR (untranslated 5′ end) contributes to their stability (Cao et al. 2017). Nevertheless, it has been shown that *Euryarcheaota* and *Crenarcheaota* predominantly possess leaderless mRNAs (Londei 2005; Babski et al. 2016). Maturation of the 5′ triphosphate end to monophosphate is thought to be the limiting step initiating 5′–3′ degradation of mRNAs (Clouet-d'Orval et al. 2018). In *Sulfolobus solfataricus*, it has been suggested that the translation factor aIF2 binds to this 5′ triphosphate end to protect the transcript (Hasenöhrl et al. 2008; Arkhipova et al. 2015).

Archaeal mRNAs appear to have significantly longer half-lives than bacterial mRNAs (Evguenieva-Hackenberg and Klug 2009, 2011). Transcript half-lives range from 7 to 57 min in *Methanococcus vannielii* (Hennigan and Reeve 1994), from 4 to 80 min in *Haloferax mediterranei* (Jäger et al. 2002) and from 2 to 20 min in *Sulfolobus solfataricus* and *Sulfolobus acidocaldarius* (Bini et al. 2002; Andersson et al. 2006). Nevertheless, it is too early to generalize, as these observations have been made in a limited number of archaea and mRNAs. It has also been shown that the half-life of certain mRNAs in *Methanosarcina acetivorans* can be adjusted according to the carbon source (Peterson et al. 2016).

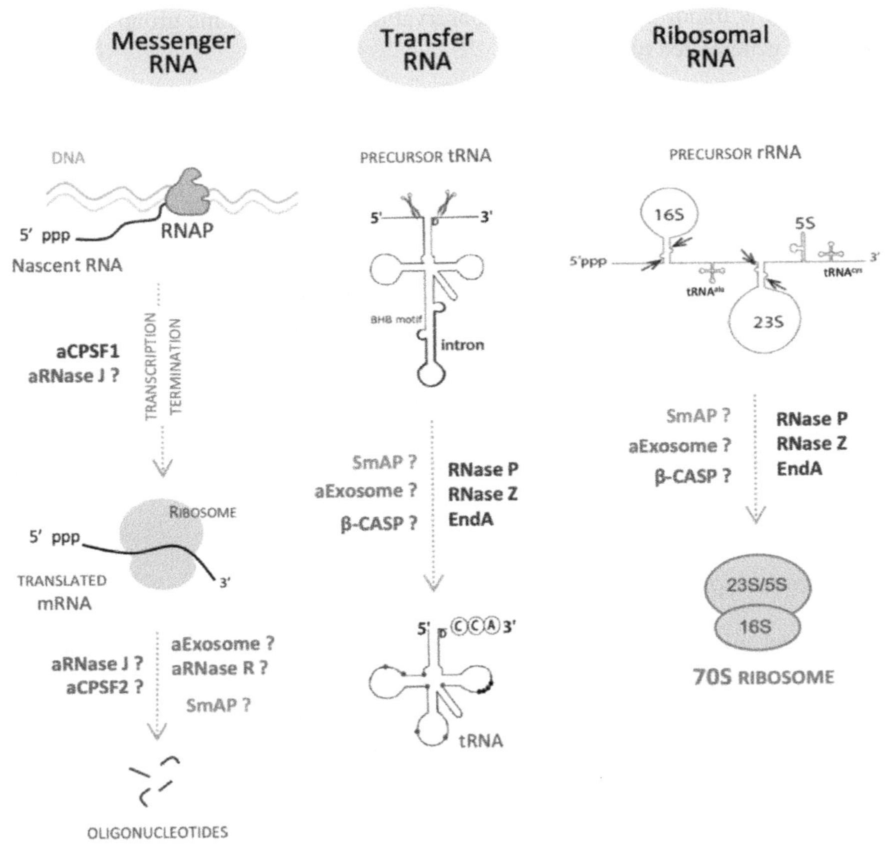

Figure 4.2. *RNA classes in archaea and their degradation and maturation enzymes. For a color version of this figure, see www.iste.co.uk/clouet/archaea2.zip*

COMMENT ON FIGURE 4.2.– *Simplified schematic representation. The single archaeal RNA polymerase transcribes all of the different types of RNA. Some RNAs are transcribed as precursors, which are then taken over by ribonucleases to become functional. Only a few steps in the maturation of tRNAs, rRNAs and crRNAs have been described. The formation of RNP C/D and H/ACA complexes for RNA modification has been better characterized. The role of these enzymes in regulating gene expression remains to be defined. 5′–3′ and 3′–5′ exoribonucleases are shown in blue and green, respectively. Endoribonucleases (including aCPSF1, which also has 5′–3′ exo activity) and RNA-binding proteins are shown in red and purple, respectively.*

4.2.2. *Transfer RNA*

Archaeal genomes contain between 36 (*Methanobrevibacter smithii*) and 61 (*Methanosarcina barkeri*) genes encoding tRNAs (Chan and Lowe 2009). These genes occur as a single transcription unit, or as an operon with other tRNAs, rRNAs or mRNAs.

tRNAs are transcribed as precursors (pre-tRNAs) (Figure 4.2). Their maturation, which requires processing at both ends of the pre-tRNA, is a highly conserved process in all three domains of life (Clouet-d'Orval et al. 2018): the 5′ and 3′ ends of tRNAs are generated by the ribonucleases RNase P and RNase Z, respectively; a CCA is added at 3′ by a tRNA nucleotidyl-transferasese. On average, 15% of archaeal tRNAs contain introns, ranging in size from 11 to 129 nucleotides (Kanai 2014). Their splicing is carried out by the EndA endonuclease, similar to the eukaryotic tRNA-splicing endonuclease (Thompson and Daniels 1988). In some archaea, including *Nanoarchaeum equitans*, tRNAs can be encoded on two separate genes, known as "split tRNAs". The two parts of the tRNA are then joined by EndA to form a mature tRNA (Randau et al. 2005; Chan et al. 2011). The tRNA biogenesis process is described in Chapter 5 of this volume.

4.2.3. *Ribosomal RNA*

In all three domains of life, rRNAs are produced from polycistronic precursors. They are matured during the complex process of ribosome assembly (Figure 4.2). Archaeal ribosomes sediment at 70S and are composed of 23S, 16S and 5S rRNAs. The majority of archaeal genomes only have one copy of the ribosomal RNA operon (*rrn*) (Clouet-d'Orval et al. 2018; Londei and Ferreira-Cerca 2021). Some archaea, such as *Methanococcus vannielii* or *Halolamina sediminis*, possess up to four copies, distributed throughout their genomes (Acinas et al. 2014).

In archaea, there are predominantly two types of *rrn* operon. In both cases, 16S and 23S rDNA are co-transcribed. In *Euryarcheaota*, 5S rDNA is found downstream of 23S rDNA. In most cases, this *rrn* operon also contains genes encoding tRNAs. In TACK, 5S rRNA is usually transcribed from a separate gene (Londei and Ferreira-Cerca 2021).

During ribosome biogenesis, rRNA precursors are taken up by RNases to become mature 23S, 16S and 5S rRNAs. Endoribonucleases, some of which are universally conserved, have been identified as involved in these processes, but many players remain to be discovered (Clouet-d'Orval et al. 2018; Qi et al. 2020; Londei and Ferreira-Cerca 2021). The regions surrounding 16S and 23S rRNA sequences contain BHB motifs that are cleaved by the endoribonuclease EndA (Schwarz et al. 2020). As a result, 16S and 23S pre-rRNAs are circular. Linearization of these circular intermediates and their final maturation are carried out by enzymes that are still largely undiscovered. Only the maturation of the 3′ end of 16S rRNA by the endoribonuclease Nob1 has been identified (Veith et al. 2012). For 5S rRNA, the mature 5′ end is generated by RNase Z, while the enzyme producing the 3′ end has yet to be identified (Hölzle et al. 2008). The processes involved in rRNA biogenesis are described in Chapter 5 of this volume.

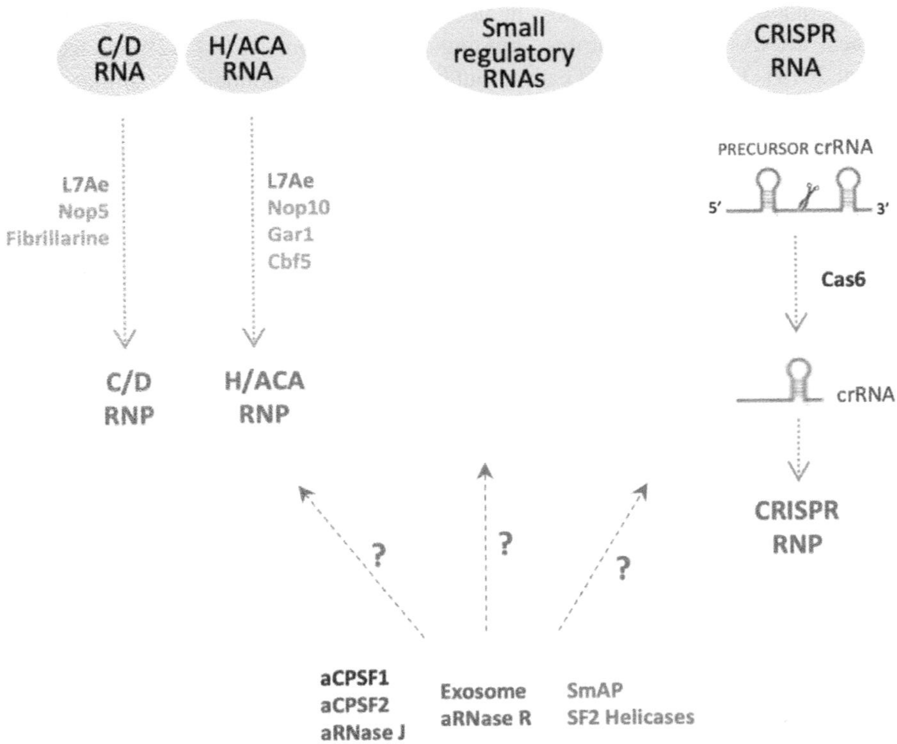

Figure 4.3. *RNA classes in archaea and their degradation and maturation enzymes (see comment on Figure 4.2). For a color version of this figure, see www.iste.co.uk/clouet/archaea2.zip*

4.2.4. *Box C/D and H/ACA guide RNA*

Box C/D RNAs associate with L7Ae, Nop5 and fibrillarin to form a ribonucleoprotein complex (RNP) that guides 2′-O-methylation of ribose on rRNAs and tRNAs (Omer et al. 2002; Breuer et al. 2021) (Figure 4.3). Box H/ACA RNAs and the proteins L7Ae, Nop10, Gar1 and Cbf5 associate to form the RNP complex that guides pseudouridylation on rRNA uridines (Charpentier et al. 2007; Liang et al. 2009; Yip et al. 2013).

In archaea, while there are numerous box C/D RNAs (Gaspin et al. 2000; Tripp et al. 2017; Breuer et al. 2021), particularly in thermophilic species, only a few box H/ACA RNAs have currently been identified (Dennis et al. 2015). These guide RNAs and the RNP complexes they form are conserved in eukaryotes and absent in bacteria. Box C/D RNAs are predominantly synthesized co-transcriptionally with neighboring genes. Maturation stages are therefore necessary to obtain their mature forms. A significant number of box C/D RNAs have been found in circular form. This is the case of box C/D RNAs, which are intronic to certain tRNAs (Bortolin et al 2003). They are excised by the endoribonuclease EndA and circularized by the ligase RtcB (Clouet-d'Orval et al. 2018; Breuer et al. 2021). These RNAs are described in detail in Chapter 6 of this volume.

4.2.5. *Small regulatory RNA*

Small regulatory RNAs, or noncoding RNAs (ncRNAs), are involved in many biological processes: metabolic regulation, adaptation to extreme conditions and stress response. As in bacteria and eukaryotes, they likely act by modifying the translatability and stability of their mRNA targets. There are three classes: intergenic ncRNAs, antisense ncRNAs and sense ncRNAs. They have been detected in several *Crenarcheaota* and *Euryarcheaota*: *Archaeoglobus fulgidus* (Tang et al. 2002); *Sulfolobus solfataricus* (Tang et al. 2005); halophiles (Soppa et al. 2009; Straub et al. 2009; Fischer et al. 2011); *Methanolobus psychrophilus* (Li et al. 2015); *Thermococcus kodakarensis* (Jäger et al. 2014); *Methanosarcina mazei* (Jäger et al. 2009); and *Pyrococcus abyssi* (Phok et al. 2011; Toffano-Nioche et al. 2013). Some are expressed differentially, depending on growth conditions (Straub et al. 2009; Jaschinski et al. 2014). Archaeal ncRNAs are considered to be of the bacterial type as they are not matured, as is the case with eukaryotic microRNAs. These RNAs are described in detail in Chapter 6 of this volume.

4.2.6. *CRISPR RNA*

CRISPR RNAs (crRNAs) are involved in the cell's immune defense against infectious genetic elements, such as plasmids, viruses or transposable elements (Figure 4.3). They are part of the CRISPR-Cas systems found in the majority of archaeal genomes (Makarova et al. 2015, 2020). To briefly summarize, CRISPR locus encode short palindromic repeats, spaced by unique sequences called spacers, that pair with foreign genetic elements. Associated Cas genes encode proteins required for the defense response, including RNA-binding proteins, nucleases, helicases or polymerases. Transcription of the CRISPR locus produces a long pre-crRNA, which is cleaved at each repeat into individual crRNAs.

During the interference phase, invasive DNA or RNA molecules that pair with crRNAs are cleaved by effector RNP complexes, composed by Cas proteins associated with individual crRNAs (Koonin et al. 2017). The new classification of CRISPR-Cas systems includes two classes, six types and 33 sub-types that differ according to their composition and function (Makarova et al. 2020). Class 1 has a multi-protein effector complex, while Class 2 is composed of a single protein. Archaea generally possess the Class 1 CRISPR system.

4.3. Ribonuclease families in archaea

Ribonucleases (RNases) are key players in RNA metabolism (Stoecklin and Mühlemann 2013). They catalyze the exoribonucleolytic or endoribonucleolytic cleavage of phosphodiester bonds. They are classified into two main groups: hydrolases and phosphorylases. Hydrolases require water molecules to catalyze the cleavage of phosphodiester bonds, while phosphorylases use inorganic phosphate (Pi) to release nucleoside 5′-diphosphates (NPD).

There are 24 ribonucleases in *Escherichia coli*, 34 in *Saccharomyces cerevisiae* and 60 in humans. They have a wide range of functions and are involved in RNA maturation, quality control and degradation. They can also act against viral genome invasions or the spread of transposable elements. This section lists ribonucleases identified in various archaeal genomes (Figure 4.4). In most cases, RNases have been identified based on their similarities with eukaryotic and bacterial RNases (Clouet-d'Orval et al. 2018) (Table 4.1).

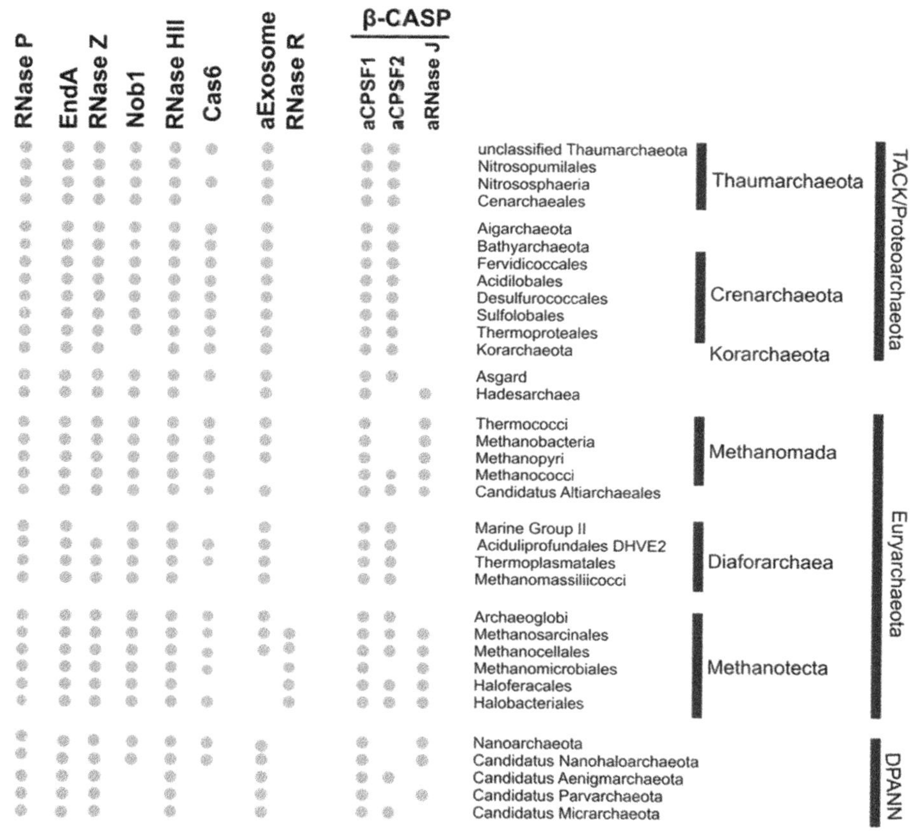

Figure 4.4. *Taxonomic distribution of ribonuclease families in archaeal phylogeny*

COMMENT ON FIGURE 4.4.– *The presence of each family in the taxonomic group is indicated by a gray circle. The classification of taxonomic groups is in line with the phylogenetic tree (Adam et al. 2017), with the exception of the position of cluster I. The DPANN group (Diapherotrites, Parvarchaeota, Aenigmarchaeota, Nanoarchaeota, Nanohaloarchaea) refers to the superphylum of extremophilic archaea (Rinke et al. 2013). The search for ribonuclease families in these different groups of archaea is described in Clouet-d'Orval et al. (2018).*

	Family		Name	Role	Bact.*	Euc.*
3-5' EXO	RNB		**RNaseR**	*n.d.*	✓	✓
3-5' EXO	PDX		**Rrp41/Rrp42**	*n.d.*	✓	✓
5-3' EXO	β-lactamase	β-CASP	**aRNase J**	*n.d.*	✓	
5-3' EXO	β-lactamase	β-CASP	**aCPSF2**	*n.d.*		✓
ENDO & 5-3' EXO	β-lactamase	β-CASP	**aCPSF1**	*n.d.*		✓
ENDO	β-lactamase	Elac	**RNase Z**	tRNA maturation	✓	✓
ENDO	RNase P		**RNase P**	tRNA maturation	✓	✓
ENDO	SEN		**EndA**	tRNA & rRNA maturation		✓
ENDO	PIN		**Nob1**	rRNA maturation	✓	✓
ENDO	Ferredoxin-like		**Cas6**	CRISPR RNA maturation	✓	
ENDO	RNase H		**HI/HII**		✓	✓

Table 4.1. *Ribonuclease families in archaea. Occurrence, activities and known functions of ribonucleases in archaea (Clouet-d'Orval et al. 2018). The occurrence of these families in eukaryotes (Euc.) and bacteria (Bact.) are described in Aravind and Koonin (2001), Condon and Putzer (2002) and Arraiano et al. (2013)*

4.3.1. *Ubiquitous endoribonucleases in archaea*

The endoribonucleolytic activities found in archaeal cells come from universal enzyme families (RNase P and RNase Z), ubiquitous families shared with eukaryotes (EndA, Nob1 and aCPSF1) and the Cas6 family of the CRISPR-Cas defense system shared with bacteria. The aCPSF1 ribonuclease, a member of the β-CASP family that is mainly composed of 5′–3′ exoribonucleases, is detailed in the corresponding section.

4.3.1.1. *RNase P*

Ribonuclease P (EC 3.1.26.5, RNase P) is a site-specific ribonuclease present in all three domains of life. It is an essential enzyme that is mainly known for its universally conserved role in the maturation of the 5′ end of transfer RNA precursors (Frank and Pace 1998; Evans et al. 2006; Jarrous and Gopalan 2010; Lai et al. 2010). It is also known to be involved in the maturation of 4.5S RNA precursor (SRP), 10Sa RNA (tmRNA) and riboswitches in bacteria, as well as in the excision of small nucleolar box C/D RNAs in eukaryotic introns (Esakova and Krasilnikov 2010; Altman 2011).

RNase P is unique from other ribonucleases in that it is a ribozyme. It forms a ribonucleoprotein complex (RNP), consisting of catalytic RNA (RPR) and RNA-binding proteins (RPP). Depending on the RPRs and RPPs of which it is composed, RNase P can form different scaffolds that enable the evolutionary history of these complexes to be traced (Lai et al. 2010; Howard et al. 2013; Klemm et al. 2016). RPRs, in particular, are structurally related and classified by substrate-binding (S-domain) and catalytic (S-domain) functional domains (Gopalan 2007). In archaea, there are three types of RPR (Samanta et al. 2016): type A, related to the bacterial type, is the most represented in archaeal phylogeny; type M, close to type A but lacking in some regions, is restricted to *Methanococci* and *Archaeoglobus*; type T, a minimal form with no S-domain, was recently identified in *Thermoproteaceae*.

Although A-type and M-type RNase P archaea contain all of the elements required for substrate recognition and catalysis, they are defective in the absence of protein. Initially, A-type and M-type RNase P archaea were described as containing at least four types of protein in heterodimer (RPP21/RPP29) and heterotetramer (POP5/RPP30) form (Samanta et al. 2016). A fifth partner, RPP38/L7Ae, has recently been shown to associate with RNase P RNA through recognition of an architectural motif called kink-turn (Cho et al. 2010; Lai et al. 2014). A structural model proposed for RNase P, from *Pyrococcus horikoshii*, shows how protein-RNA interactions would be important for making the archaeal RNase P active (Kimura and Kakuta 2012).

4.3.1.2. *EndA*

Endonuclease A (EC 4.6.1.16, EndA) is an endoribonuclease that was initially identified in eukaryotes as being involved in RNA splicing. In archaea, EndA is a strictly conserved multimeric enzyme with a variable subunit composition. This nuclease, which recognizes and cleaves specific structural motifs, known as bulge–helix–bulge (BHB) motifs, has been shown to be involved in the critical stages of maturation of ribosomal RNA and certain intron-containing transfer RNAs (Thompson and Daniels 1990; Marck and Grosjean 2003).

There are four types of EndA archaeal enzymes, which are classified, according to their compositions, into subunits and have a distinct taxonomic distribution (Tocchini-Valentini et al. 2005; Calvin and Li 2008; Clouet-d'Orval et al. 2018). The homotetrameric α4 form is mainly found in the *Euryarchaea*, but also, interestingly, in the Asgard, the archaeal group most closely related to eukaryotes (Zaremba-Niedzwiedzka et al. 2017). Another part of the *Euryarchaea* encodes the homodimeric α2 form instead. In TACK, EndA mainly comes in the heterotetrameric α2β2 form. The fourth group is an ε2 homodimer that is restricted to the ARMAN-1 and ARMAN-2 (Archaeal Richmond Mine Acidophilic Nanoorganisms) groups of archaea (Fujishima et al. 2011; Hirata et al. 2012). Comparative sequence analysis shows strong sequence similarity between the archaeal α and β subunits and the catalytic Sen2 and Sen34 subunits of eukaryotic EndA. This suggests that eukaryotic EndA evolved from archaeal EndA, with the subsequent acquisition of the accessory subunits Sen15 and Sen34 (Lopes et al. 2015).

The structures of the different EndA types have been solved and are essentially identical. The subunits possess three conserved catalytic residues, and their respective assemblies lead to the formation of two intron-cleavage sites at the active site; while all four EndA types recognize the canonical BHB motif, which comprises two three-nucleotide bulges, separated by a four-nucleotide helix located at the exon-intron boundary of transfer RNA precursors. Only the α2β2 and ε2 types can effectively eliminate introns with noncanonical or relaxed BHB motifs (Lykke-Andersen and Garrett 1997; Li et al. 1998; Li and Abelson 2000).

4.3.1.3. *RNase Z*

Endoribonuclease Z (EC 3.1.26.11, RNase Z) is a member of the enzyme metallo-β-lactamase (MβL) superfamily (Aravind 1999). MβLs, which share a low level of sequence identity, are characterized by an αβ/βα-fold and five motifs that coordinate two Zn^{2+} ions and are responsible for the hydrolysis reaction. RNase Z consists of the MβL domain, where a unique domain, known as an exosite, is inserted, forming a flexible arm involved in the attachment of its substrate (Ishii et al. 2005, 2007; Redko et al. 2007).

RNase Z, which is conserved in all three domains of life (also named Elac in eukaryotes) (Späth et al. 2008), catalyzes the maturation of the 3′ end of transfer RNA precursors to generate a 3′-OH (Ishii et al. 2005; Li de la Sierra-Gallay et al. 2006; Zhang and Ferré-D'Amaré 2016). The universal form of RNase Z is a short form of 250–350 amino acids (aa). Eukaryotes also possess a long form (750–900 aa), probably the result of gene duplication (Vogel et al. 2005).

The archaeal RNase Z is distributed in all archaeal genomes, with the exception of *Nanoarchaeota Nanopusillus acidilobi* (Clouet-d'Orval et al. 2018). The enzymatic activities of RNase Z from *Methanocaldococcus jannaschii*, *Pyrococcus furiosus*, *Haloferax volcanii*, *Pyrobaculum aerophilum* and *Thermoplasma acidophilum* were characterized. These enzymes only slightly differ in their biochemical requirements, cleavage site position and substrate specificity (Schierling et al. 2002; Hölzle et al. 2008; Späth et al. 2008).

4.3.1.4. *Nob1*

Archaea possess a Nob1 endoribonuclease (EC 3.1.) similar to the eukaryotic ribosome assembly factor Nob1, which generates the mature 3′ end of 18S ribosomal RNA at the D-site (Veith et al. 2012; Clouet-d'Orval et al. 2018). In vitro, Nob1 from *Pyrococcus horikoshii* (*Pho*) efficiently cleaves RNA substrates with a D-site in the presence of manganese. The structure of *Pho*-Nob1 was resolved by nuclear magnetic resonance (NMR) spectroscopy. As in eukaryotes, a PIN domain, characteristic of many nucleases, and a zinc ribbon finger domain are linked by a flexible segment. However, unlike *Pho*-Nob1, eukaryotic Nob1 has a long C-terminal tail and contains a large insertion in its PIN domain.

Nob1 is conserved in the majority of archaeal genomes, with the exception of *Korarchaea* and some DPANNs, whose genomes are incomplete. Moreover, in some *Crenarchaeota*, a shorter C-terminal domain has been observed. This is the case of *Thermoproteales* (13 out of 17 genomes), *Acidilobales* (2/2), *Fervidicoccales* (1/1) and some *Desulfurococcales* (3/15).

4.3.1.5. *RNase HII*

Endoribonuclease HII (EC 3.1.26.4, RNase HII) is a ubiquitous nuclease in archaea. It belongs to the RNase H family, which is conserved across the three domains of life (Clouet-d'Orval et al. 2018). This family of proteins is involved in the elimination of DNA/RNA hybrids created in the genome during processes such as DNA replication, which requires the synthesis of Okazaki fragments, and DNA repair with the excision of a misincorporated ribonucleotide (Kochiwa et al. 2007; Tadokoro and Kanaya 2009). In order to achieve this, these enzymes specifically hydrolyze the P-O3′ phosphodiester bonds of the RNA strand that is hybridized to a complementary strand of DNA. In vitro and in the presence of manganese ions, RNase HII from *Pyrococcus furiosus* also appears capable of digesting RNA/RNA homo-duplexes (Kitamura et al. 2010).

There are two RNase H families, RNase HI and RNase HII, which are differentiated on the basis of their amino acid sequences (Kochiwa et al. 2007). The RNase HII gene is conserved in all complete archaeal genomes, with the exception of *Nitrosopumilus sediminis* and *N. koreensis AR1*. In contrast, the RNase HI gene has a very low occurrence and sporadic distribution. It is found together with RNase HII in the genomes of *Halobacterium sp. NRC-1*, *Sulfolobus tokodaii* and *Pyrobaculum aerophilum*. RNase HII therefore plays an essential role in archaea and is not redundant with RNase HI.

The RNase HII structures of *Methanococcus jannaschii* (Lai et al. 2000), *Thermococcus kodakarensis* (Muroya et al. 2001) and *Archaeoglobus fulgidus* (Bubeck et al. 2011) have been solved. These structures share the folding motif known as RNase H-fold and the same organization of the four acidic residues of the active site. Interestingly, the RNase H-fold is also found in the PIWI domain of the Ago (argonaute-like) protein from *Pyrococcus furiosus*. This suggests an evolutionary relationship between RNases H and Ago proteins in prokaryotes.

4.3.1.6. *Cas6*

Defense systems based on CRISPR RNAs and their associated CAS proteins are present in many archaeal genomes (Vestergaard et al. 2014; Clouet-d'Orval et al. 2018). They associate to form ribonucleoprotein complexes that target exogenous nucleic acid sequences. The CRISPR RNAs that serve as guides are derived from long precursors and are matured by endoribonucleases. Cas6 is one of the ribonucleases required for the maturation of class 1 crRNAs (Hale et al. 2008, 2009; Makarova et al. 2020). Cas6 has two conserved motifs: a ferredoxin domain, or RRM (RNA recognition motif), and a C-terminal glycine-rich motif. It is a metal-independent nuclease, whose predicted active site of the enzyme shows similarities with the active site of the archaeal endoribonuclease EndA (Hochstrasser and Doudna 2015).

4.3.2. *3′–5′ exoribonucleases*

In archaea, 3′–5′ exoribonucleolytic activity is predominantly carried out by an RNA exosome, similar to that of eukaryotes. In the few genomes that do not encode the subunits of this exosome, an *rnr* gene is found that encodes the bacterial RNase R (Portnoy and Schuster 2006). Only the *Methanococci* group appears to lack enzymes with 3′–5′ exoribonucleolytic activity (Clouet-d'Orval et al. 2018). It is perhaps compensated for by 5′–3′ exoribonucleolytic or endoribonucleolytic degradation machineries. It is also possible that this seemingly missing activity is carried by an archaea-specific enzyme that remains to be discovered.

4.3.2.1. *aExosome*

The PH-domain ribonuclease (PH RNase) superfamily includes the bacterial Polynucleotide Phosphorylase (PNPase), as well as the eukaryotic and archaeal RNA exosomes. These enzymes are involved in the 3′–5′ degradation of all types of cellular RNA (Lin-Chao et al. 2007).

The archaeal RNA exosome (EC 3.1.13.1, aExosome) was first isolated from *Sulfolobus solfataricus* (Evguenieva-Hackenberg et al. 2003). It has mainly been studied in *Sulfolobales* and *Thermococcales*, using biochemical and structural approaches (Evguenieva-Hackenberg et al. 2014). The aExosome is composed of a catalytic core and a CAP. The catalytic core, in barrel form, is formed by an Rrp41/Rrp42 dimer trimer The catalytic site is carried by the Rrp41 subunit (Lorentzen et al. 2005, 2007). The exosome CAP is a trimer of RNA-binding proteins Rrp4 (S1 and KH domains) and Csl4 (S1 and ZN zinc finger domains), whose stoichiometry in vivo is unknown (Ramos et al. 2006; Witharana et al. 2012).

In the presence of inorganic phosphate, the aExosome catalyzes phosphorolytic degradation of single-stranded RNA, again from its 3′ end, to release nucleoside 5′-diphosphates (NDPs). It is also capable of the opposite reaction, using NDPs to synthesize heteropolymers, such as polyA tails, at the 3′ ends of RNAs (Evguenieva-Hackenberg et al. 2017). Structurally, this nine-subunit complex resembles the bacterial PNPase and the eukaryotic exosome core. In terms of catalytic activity, the aExosome is closer to the bacterial PNPase.

In humans and yeast, the exosome core is catalytically inactive and comprises a hexameric ring of PH RNases (Rrp41, Rrp42, Mtr3, Rrp43, Rrp46 and Rrp45), and RNA-binding proteins Rrp4, Rrp40 and Csl4. The inactive core of the exosome associates with two hydrolytic ribonucleases, Rrp44/Dis3 and Rrp6, which give it its ability to degrade RNA (Lebreton and Séraphin 2008; Schaeffer et al. 2009). In bacteria, PNPase is an enzyme homotrimer, with each subunit comprising two PH domains and RNA-binding domains KH and S1. Only one of the PH domains of each subunit appears to carry the catalytic activity of the PNPase. Beyond this common architecture, PNPase shares the same phosphorolytic and synthetic activities as the aExosome (Briani et al. 2016).

In vitro, the "capped" aExosome degrades RNA more efficiently than the hexameric barrel alone (Büttner et al. 2005; Evguenieva-Hackenberg et al. 2008). In *Sulfolobus solfataricus*, the Csl4 subunit enables the interaction of the aExosome with the aDnaG protein, which, in turn, is capable of interacting with single-stranded RNA-binding smAP proteins (Witharana et al. 2012; Hou et al. 2013). The exosome's high specificity for poly(A) RNAs is enabled by the Rrp4 and aDnaG subunits (Roppelt et al. 2010; Hou et al. 2013). The aExosome, through Rrp4, also interacts with Nop5 (eukaryotic Nop58 homolog), a protein that is part of the RNA methylation box C/D ribonucleoprotein complex (Gauernack et al. 2017). An interaction between Csl4 and the exoribonuclease aRNase J has been characterized in *Pyrococcus abyssi* (Phung et al. 2020).

Sequencing of RNAs co-immunoprecipitated (iCLIP) with the *S. solfataricus* aExosome has, for the first time, identified its major role in RNA metabolism in archaea (Bathke et al. 2020). The Sso-aExosome interacts with numerous messenger RNAs, suggesting a fundamental role in mRNA degradation. Enrichment of the 5′ and 3′ ends of mRNAs suggests an approximation of the ends during degradation that could prevent translation of degrading RNAs. Noncoding RNAs and circular RNAs (precursors of 16S and 23S ribosomal RNAs) were also found in complex with Sso-aExosome. Recruitment of these RNAs takes place at short poly(A) sequences that may be internal to the RNAs. Moreover, a significant number of these RNAs, containing additional poly(A) tails, have also been identified.

4.3.2.2. *aRNase R*

Exoribonuclease R (EC 3.1.13.1, aRNases R) belongs to the RNase II/RNB family, which is characterized by an RNB domain responsible for catalytic activity. In archaea, there are three types of RNase R, arbitrarily named (a), (b) and (c), which bear no resemblance to bacterial RNase R. The taxonomic distribution of RNase R is mostly restricted to *Haloarchaea* and a few *Methanococcales* (Clouet-d'Orval et al. 2018). Most of these organisms do not possess an RNA exosome, and RNase R is the only source of 3′–5′ exoribonuclease activity. In halophiles, RNase R(a) is restricted to the RNB domain. The RNase R(b) and RNase R(c) forms, which contain an accessory C-terminal domain with RNA-binding properties, are only found in a few *Methanococcales*. It has been suggested that these different forms of RNase R are likely the result of horizontal gene transfer from bacterial genomes.

In bacteria, RNase R degrades structured RNAs from their 3′ ends to their 5′ ends, and is responsible for important functions in the cell, from RNA quality control to virulence (Arraiano et al. 2013).

In *Haloferax volcanii*, the gene encoding RNase R(a) appears to be essential (Portnoy and Schuster 2006; Matos et al. 2012). This may be explained by the fact that this is the only protein appears to possess 3′–5′ exoribonuclease activity in halophiles. Although the catalytic activity of this protein has been determined, nothing is known about its role in cellular RNA degradation pathways.

4.3.3. *The β-CASP ribonuclease and 5′–3′ exoribonuclease families are widespread in archaea*

The β-CASP nuclease family includes enzymes for RNA cleavage (CPSF73 and RNase J) and DNA repair (Artemis, SNM1 and PSO2) (Dominski et al. 2013). It is a member of the MβL superfamily, enzymes found in all three domains of life (Dominski 2007). Members of this family contain an MβL domain (MβL see RNase Z) into which is inserted a β-CASP domain that is formed by a central sheet of six parallel β-strands, surrounded by five α-helices. The catalytic site pocket is formed at the interface of the MβL and β-CASP domains, which coordinate two divalent cations (Zn^{2+}, Mn^{2+} or Mg^{2+}). It is capable of carrying out two types of 5′–3′ exoribonucleolytic and endoribonucleolytic reactions, catalyzed by the same enzyme site (Li de la Sierra-Gallay et al. 2008; Dorléans et al. 2011; Newman et al. 2011). This dual enzymatic function within the same catalytic site is not yet well understood. It could be generated by the binding of another factor or via association in a multi-protein complex.

Among archaeal β-CASP enzymes, two groups have emerged as playing fundamental roles in RNA metabolism (Clouet-d'Orval et al. 2010; Phung et al. 2013). One is linked to the eukaryotic CPSF73 protein, and the other to the bacterial RNase J protein.

4.3.3.1. *The CPSF group*

In archaea, there are numerous CPSF group members that are classified into three ortholog groups: aCPSF1, aCPSF1b and aCPSF2 (Phung et al. 2013; Clouet-d'Orval et al. 2015). They are homologous to the eukaryotic factor CPSF73 (Dominski et al. 2013). This ubiquitous factor is part of the CPSF (cleavage and polyadenylation specificity factor) multiprotein complex, which is involved in 3′-end processing and polyadenylation of all RNA polymerase II messenger RNA transcripts. CPSF73 is an endoribonuclease that specifically recognizes the CA motif, located 20–30 nucleotides upstream of the polyadenylation consensus site. In metazoans, CPSF73 also has 5′–3′ exoribonucleolytic activity and is involved in the maturation of histone-encoding messenger RNA precursors. In bacteria, there are sporadically distributed CPSF homologs, whose function has yet to be described (Dominski et al. 2013).

4.3.3.2. *aCPSF1*

In addition to MβL and β-CASP domains, the ribonuclease aCPSF1 (EC 3.1.4.1) possesses an N-terminal accessory domain consisting of two nucleic acid-binding KH domains and a C-terminal protein dimerization motif. Crystallographic structures of aCPSF1 from *Pyrococcus horikoshii*, *Methanosarcina mazeii* and *Methanolobus thermautotrophicus* show a three-part architecture, in which the KH domains are connected to the central core through a long linker (Nishida et al. 2010; Mir-Montazeri et al. 2011; Silva et al. 2011). aCPSF1 has a dual activity in *Thermococcales* and *Methanococcales*: it is able to internally cleave single-stranded RNA at a CA site and possesses 5′–3′ exoribonucleolytic activity that is restricted to monophosphorylated 5′ RNA substrates (Levy et al. 2011; Phung et al. 2013). Its activity is comparable to that of CPSF73 in eukaryotes. The presence of aCPSF1 across archaeal genomes (Clouet-d'Orval et al. 2018) suggests that its activity is central to conserved metabolic pathways. It has recently been suggested that it is involved in transcription termination in *Thermococcus kodakarensis*. As a result of this function, it is also named FttA, for "Factor that terminates transcription in Archaea" (Sanders et al. 2020).

aCPSF1b represents a small group of β-CASP ribonucleases, restricted to the *Methanococci* and *Korachaeota* group, and is closely related to aCPSF1 (Phung et al. 2013; Clouet-d'Orval et al. 2018). Its 5′–3′ exoribonucleolytic activity has been shown in *Methanococcus jannaschii*, but no endonucleolytic activity has been observed (Levy et al. 2011). Members of this group lack the accessory N-terminal domain and, to date, no aCPSF1b structure has been solved.

4.3.3.3. *aCPSF2*

The ribonuclease aCPSF2 (EC 3.1.4.1) is a less conserved cluster in archaea compared to the aCPSF1 cluster (Phung et al. 2013; Clouet-d'Orval et al. 2018). The members of this cluster, found in TACK/*Proteoarchaeota* and sporadically in *Euryarchaeota*, are restricted to the MβL and β-CASP domains, except in *Sulfolobus solfataricus*, which possesses an N-terminal extension. Currently, no aCPSF2 structure has been solved. The aCPSF2 proteins of *Sulfolobus solfataricus* and *Sulfolobus acidocaldarius* possess 5′–3′ exoribonucleic activity that depends on the presence of Mg^{2+} ions (Levy et al. 2011; Märtens et al. 2013). In *S. solfataricus*, the gene encoding aCPSF2 can be deleted without affecting cell doubling time.

4.3.3.4. *aRNase J*

Ribonuclease J (EC 3.1.13.B1, RNase J) is present in bacteria, archaea and chloroplasts (Dominski et al. 2013). In bacteria, bRNase J is present in over half of all species, particularly Gram-positive bacteria. It plays a key role in ribosomal RNA maturation and messenger RNA degradation (Mathy et al. 2007; Linder et al. 2014; Durand and Condon 2018). In addition to the MβL and β-CASP domains, bRNase J generally possesses a C-terminal extension (CTD). In *Bacillus subtilis*, the CTD domain is notably involved in the dimerization of the enzyme (Li de la Sierra-Gallay et al. 2008). bRNase J possesses dual 5′–3′ exoribonucleolytic and endoribonucleolytic activity. Most Firmicutes possess two paralogs, bRNase J1 and bRNase J2, which are very similar in their primer sequences, with no apparent difference in their structural domains (Even et al. 2005). bRNase J1 and J2 can merge to form heterocomplexes (dimer or tetramer). The formation of these complexes has significant consequences for 5′–3′ exoribonucleolytic activity, as well as for endonucleolytic cleavage site specificity.

In chloroplasts, the RNase J protein appears to be involved in RNA quality control, eliminating long antisense transcripts and degrading the 5′ ends of RNAs (Sharwood et al. 2011; Hotto et al. 2020). It contains an N-terminal transit peptide, and a C-terminal GT-1-type DNA-binding domain (except in algae), whose role is yet to be described.

Archaeal RNase J, or aRNase J, is restricted to *Euryarchaeota*, with the exception of the *Diafoarchaea* and Archeoglobi groups (Phung et al. 2013; Clouet-d'Orval et al. 2018). It is restricted to the MβL and β-CASP domains but has two specific insertions of 9 and 12 amino acids in the β-CASP domain (Loop 1), as well as in the N-terminal part of the MβL domain (Loop 2), respectively (Clouet-d'Orval et al. 2010). In vitro, recombinant aRNase J proteins from *Thermococcales* and *Methanococci* possess highly processive 5′–3′ exoribonucleolytic activity, no detectable endoribonucleolytic activity under the conditions tested, and greatly reduced activity in degrading 5′-triphosphorylated RNA substrates (Clouet-d'Orval et al. 2010; Levy et al. 2011).

Structure resolution of aRNase J from *Methanolobus psychrophilus* R15 shows that Loops 1 and 2 interact with the RNA molecule (Zheng et al. 2017). They are also important for the ribonucleolytic activity of aRNase J from *Pyrococcus abyssi* (Clouet-d'Orval et al. 2010). Biochemical, proteomic and phylogenomic analyses suggest that Pab-aRNase J participates in mRNA degradation pathways near the ribosome, in coordination with the ASH-Ski2 helicase and/or the aExosome (Phung et al. 2020).

4.4. Ubiquitous RNA-binding proteins in archaea

Ribonucleases are not the only players in RNA metabolism. RNA-binding proteins stabilize and structure RNAs and are important in shaping the ribonucleoprotein complexes in which RNases operate.

4.4.1. *Sm-like proteins*

Members of the Sm-like/Hfq proteins form a superfamily of RNA-binding proteins that have ancient origin (Mura et al. 2013). They contain a common bipartite sequence, known as the Sm domain, consisting of two Sm motifs linked by a region of variable length and motif. While the proteins are not similar in the way of primary structure, they do share a tertiary structure and a multimeric quaternary architecture that favors interactions with partner molecules. Another conserved feature of Sm-like proteins is their preference for single-stranded, uracil-rich RNAs.

In archaea, Sm-like proteins have been named SmAP for Sm-like archaeal proteins. Archaea often possess two SmAP paralogs, SmAP1 and SmAP2. In *Pyrococcales* and halophiles (*Euryarchaeota*), a single Sm-like protein has been identified, orthologous to SmAP1. There is a high degree of sequence similarity between members of the SmAP1 family, with up to 60% sequence identity. The level of identity falls to 30% or less between SmAP1 and SmAP2 proteins from the same organism. *Crenarcheota*, such as *Sulfolobus solfataricus*, *S. tokodaii* and *Pyrobaculum aerophilum*, may possess an additional Sm-like paralog, SmAP3. SmAP3 proteins are characterized by a longer C-terminal end. Lastly, *Methanococcus jannaschii* contains an Hfq-like protein instead of a conventional Sm-like protein.

The function of SmAP proteins in archaeal cells is largely unknown. In *Sulfolobus solfataricus*, a link exists between SmAP1-2 and the aExosome through an interaction between SmAP1-2 and the aDnaG subunit. SmAPs are thought to be required for efficient polyadenylation of RNA substrates (Märtens et al. 2017). Furthermore, co-immunoprecipitation experiments show that Sso-SmAP1 would preferentially interact with intron-containing RNAs, whereas Sso-SmAP2 would predominantly bind to messenger RNAs and could be involved in their stability or degradation (Märtens et al. 2015).

4.4.2. *The ribosomal protein L7Ae*

The archaeal L7Ae protein, which belongs to the L7Ae/L30 protein family, is homologous to the human 15.5kD protein and the small nucleolar RNP proteins of yeast Snu13 (Kuhn et al. 2002; Rozhdestvensky et al. 2003). It is not conserved in bacteria. These proteins have the ability to bind to the K-turn RNA motif, particularly those of the small box C/D and H/ACA RNAs, to form ribonucleoprotein RNA-modification complexes (Mao et al. 1999; Ban et al. 2000;

Kuhn et al. 2002). Unlike 15.5kD/Snu13 proteins, L7Ae also has the ability to recognize K-loop motifs (Vidovic et al. 2000; Hamma and Ferré-D'Amaré 2004; Nolivos et al. 2005; Li and Ye 2006). In archaea, L7Ae is also part of the large ribosome subunit and RNase P complex (Kouzuma et al. 2003). It is also capable of binding small noncoding RNAs, whose functions are yet to be determined (Phok et al. 2011).

4.4.3. *Helicases of the SF2 superfamily*

Helicases are proteins that unwind nucleic acids and remodel protein-nucleic acid complexes (Jankowsky and Fairman 2007; Fairman-Williams et al. 2010). They are involved in a wide range of cellular mechanisms. RNA helicases are fundamental in their ability to orchestrate transcription, RNA processing, ribosome biogenesis, translation and RNA degradation (Khemici and Linder 2018). Superfamily 2, or SF2, is the largest and most diverse group of helicases, with more than ten families. All three domains of life encode SF2 helicases in their genomes. SF2 members are non-hexameric helicases that share a conserved helicase core, composed of two RecA domains with nine characteristic motifs, and which often contain N-terminal and/or C-terminal accessory domains involved in regulating their activities. The helicase core provides the active site for ATP hydrolysis, binds nucleic acid and exerts basal unwinding activity.

In archaea, few RNA helicases have been identified and studied. Much more is known about DNA helicases, which are involved in DNA replication and repair (Chamieh et al. 2016). In the following, we present three families of helicases that have been identified as mediating processes involving RNA strands or ribonucleotides.

4.4.3.1. *ASH-Ski2*

The Ski2-like family includes RNA and DNA helicases (Johnson and Jackson 2013; Chamieh et al. 2016; Khemici and Linder 2018). They are conserved in eukaryotic and archaeal genomes. In addition to the conserved helicase core, the central SF2 domain, which contains 12 conserved motifs (Fairman-Williams et al. 2010), Ski2-type helicases feature a winged-helix (WH) domain and a Ratchet domain. They may also contain accessory domains attached to or inserted in the reserved domains, which modulate their activities and functions. As helicases, Ski2-like process RNAs with 3′–5′ polarity. They generally function within larger complexes and, in many cases, appear to act as a platform for the assembly of these complexes. The translocation mechanism of this helicase family along RNA or DNA molecules has been proposed on the basis of the solved structures of the archaeal

DNA helicase Hel308, from *Archeoglobus fulgidus* (Büttner et al. 2007; Johnson and Jackson 2013).

Our knowledge of Ski2-like helicases is largely based on eukaryotic model organisms. Ski2-like helicases involved in RNA metabolism include Mtr4, Ski2, Brr2 and Slh1/Rqt2. The Mtr4 and Ski2 RNA helicases are associated with the nuclear or cytoplasmic RNA exosome, respectively, and promote RNA degradation. Mtr4 acts within the TRAMP, PAXT and NEXT complexes to target specific RNA substrates (Bernstein et al. 2008; Meola et al. 2016; Patrick et al. 2017; Weick et al. 2018). Ski2 is part of the Ski complex that channels ribosome-bound RNA to the RNA exosome in order to trigger mRNA degradation in the cytoplasm (Halbach et al. 2012, 2013; Schmidt et al. 2016; Kögel et al. 2020; Zinoviev et al. 2020). The Ski2-like helicases, Brr2 and Slh1/Rqt2, are particularly large, with duplicated Ski2-like domains (Johnson and Jackson 2013). Brr2 is essential for spliceosome activation (Zhang et al. 2015; Vester et al. 2020) and Slh1/Rqt2 is involved in the ribosome quality control (RQC) pathway, as part of the RQC-trigger (RQT) complex (Matsuo et al. 2017; Sitron et al. 2017).

In archaea, the best-known member of the Ski2-like family is the Hel308/Hjm DNA helicase, which is involved in DNA repair pathways (Marini et al. 2002; Guy and Bolt 2005; Tafel et al. 2011). More recently, a new archaea-specific Ski2-like group has been identified as being potentially involved in RNA processing pathways (Phung et al. 2020). This factor, named ASH-Ski2 for archaeal-specific helicase (Chamieh et al. 2016; Phung et al. 2020), is also known as ETA (euryarchaeal termination activity) (Walker et al. 2017). In *Thermococcales*, it has been suggested that ASH-Ski2 interacts with aRNase J to form an RNA degradation complex, whose function remains to be determined (Phung et al. 2020). It has also been proposed that ETA is part of the transcription elongation complex and participates in the regulation of transcription termination (Zhai et al. 2018; Maier and Marchfelder 2019). Many questions remain open about the role of Ski2-type helicases in archaeal RNA metabolism.

4.4.3.2. *DEAD-box*

DEAD-box helicases are present in all three domains of life and are key players in RNA metabolism (Linder and Jankowsky 2011; Redder et al. 2015; Chamieh et al. 2016). They are characterized by the presence of a specific amino acid motif, Asp-Glu-Ala-Asp (DEAD). Despite strong conservation between proteins in this family, they participate in many different processes. DEAD-box helicases have been poorly characterized in archaea. They are present in most *Euryarchaeota*, with the exception of *Pyrococcales* and *Halobacteriales*. In *Crenarchaeota*, they are only found in *Sulfolobales*. Studies in *Methanococcoides burtonii* and *Thermococcus kodakarensis* have shown that these DEAD-box helicases are induced by cold stress (Lim et al. 2000; Shimada et al. 2009).

4.4.3.3. *aLhr2*

Among SF2 helicases, large helicase-related (Lhr) proteins are poorly characterized. They are found in some bacteria but are ubiquitous in archaea (Chamieh et al. 2016; Hajj et al. 2021). To date, no homologs of Lhr proteins have been identified in eukaryotes. Lhr proteins are defined by the unique organization of their domains. The "Lhr core" is composed of an SF2 helicase core, a winged-helix motif and a Lhr-specific domain 4 (Ejaz and Shuman 2018).

There are three archaea-specific Lhr groups (Hajj et al. 2021): aLhr1, aLhr2 and aLhr3. The aLhr3 group, which appears to be restricted to *Sulfolobales*, has a deteriorated domain 4. The aLhr1 and aLhr2 proteins, however, are widespread in archaeal genomes and are often present together. While aLhr2 proteins are restricted to the "Lhr core", aLhr1 proteins have an additional C-terminal domain containing a conserved zinc finger-like motif. The aLhr2 group appears to share a common ancestor with the bacterial bLhr-HTH group, while the aLhrA group does not appear to share a link with any of the bacterial groups.

The cellular functions of aLhr proteins have been little studied and remain unclear. Only members of the aLhr2 group have been characterized. In vitro, aLhr2 proteins from *Sulfolobus solfataricus* (Hel112), *S. acidocaldarius* (RecQ-like) (De Felice et al. 2007) and *Methanothermobacter thermautotrophicus* (Buckley et al. 2020) have been defined as DNA helicases with 3′–5′ polarity and ATP-dependent activity. The aLhr2 protein from *Pyrococcus abyssi* has been characterized as a DNA/RNA helicase by its ability to unwind a DNA:RNA hybrid (Hajj et al. 2021). This ability is shared by Mth-aLhr2 (Buckley et al. 2020).

In vivo, Sso- aLhr2 and Mth-aLhr2 proteins have been shown to be potentially involved in DNA repair (De Falco et al. 2018; Buckley et al. 2020). The Pab-aLhr2 protein has been detected in the interaction networks of proteins involved in DNA replication and repair (RPA complex) (Pluchon et al. 2013), as well as in the interaction network of the RNA helicase ASH-Ski2 (Phung et al. 2020). Further studies are needed to fully understand the role of this enzyme, which seems able to mediate reactions involving an RNA strand and to involve enzymes regulating RNA stability. In particular, it has been shown in recent years that RNA can serve as a template for DNA repair. Similarly, RNA-binding proteins have been identified as players in the response to DNA damage.

4.5. Conclusion

Numerous players in archaeal RNA metabolism have been identified, most often on the basis of their homologies with their eukaryotic and bacterial counterparts. To date, around ten ribonucleases have been identified, some of which have highly conserved functions throughout evolution, such as ribosomal and transfer RNA maturation. A number of RNA-binding proteins have also been recorded. Nevertheless, for many, the cellular functions of these proteins in archaeal cells remain to be discovered. Several key steps in rRNA maturation are still far from being understood. The molecular basis of messenger RNA turnover, which shapes the cell proteome, is still largely unknown. The challenge now is to identify the pathways that control the lifecycle of messenger RNAs. We also expect to discover other ribonucleases and RNA-modifying enzymes, particularly those specific to archaea. It should also be noted that archaea species have very diverse ecological niches, and may have evolved specific pathways required to maintain the integrity of their RNAs, particularly under extreme conditions. These discoveries, taken together with what is known about eukaryotes and bacteria, will broaden our understanding of the evolution of the molecular machine involved in RNA metabolism.

4.6. References

Acinas, S.G., Marcelino, L.A., Klepac-Ceraj, V., Polz, M.F. (2004). Divergence and redundancy of 16S rRNA sequences in genomes with multiple rrn operons. *J. Bacteriol.*, 186(9), 2629–2635.

Adam, P.S., Borrel, G., Brochier-Armanet, C., Gribaldo, S. (2017). The growing tree of Archaea: New perspectives on their diversity, evolution and ecology. *ISME J.*, 11(11), 2407–2425.

Altman, S. (2011). Ribonuclease P. *Philosophical Transactions of the Royal Society B*, 366(1580), 2936–2941.

Andersson, A.F., Lundgren, M., Eriksson, S., Rosenlund, M., Bernander, R., Nilsson, P. (2006). Global analysis of mRNA stability in the archaeon Sulfolobus. *Genome Biol.*, 7(10), R99.

Aravind, L. (1999). An evolutionary classification of the metallo-beta-lactamase fold proteins. *Silico. Biol.*, 1(2), 69–91.

Aravind, L. and Koonin, E.V. (2001). A natural classification of ribonucleases. *Methods Enzymol.*, 341, 3–28.

Arkhipova, V., Stolboushkina, E., Kravchenko, O., Kljashtorny, V., Gabdulkhakov, A., Garber, M., Nikonov, S., Märtens, B., Bläsi, U., Nikonov, O. (2015). Binding of the 5'-triphosphate end of mRNA to the γ-subunit of translation initiation factor 2 of the Crenarchaeon *Sulfolobus solfataricus*. *J. Mol. Biol.*, 427(19), 3086–3095. doi: 10.1016/j.jmb.2015.07.020.

Arraiano, C.M., Mauxion, F., Viegas, S.C., Matos, R.G., Séraphin, B. (2013). Intracellular ribonucleases involved in transcript processing and decay: Precision tools for RNA. *Biochim. Biophys. Acta.*, 1829(6/7), 491–513.

Babski, J., Haas, K.A., Näther-Schindler, D., Pfeiffer, F., Förstner, K.U., Hammelmann, M., Hilker, R., Becker, A., Sharma, C.M., Marchfelder, A., Soppa, J. (2016). Genome-wide identification of transcriptional start sites in the haloarchaeon *Haloferax volcanii* based on differential RNA-Seq (dRNA-Seq). *BMC Genomics*, 17(1), 629.

Ban, N., Nissen, P., Hansen, J., Moore, P.B., Steitz, T.A. (2000). The complete atomic structure of the large ribosomal subunit at 2.4 A resolution. *Science*, 289(5481), 905–920.

Bathke, J., Gauernack, A.S., Rupp, O., Weber, L., Preusser, C., Lechner, M., Rossbach, O., Goesmann, A., Evguenieva-Hackenberg, E., Klug, G. (2020). iCLIP analysis of RNA substrates of the archaeal exosome. *BMC Genomics*, 21(1), 797.

Bernstein, J., Patterson, D.N., Wilson, G.M., Toth, E.A. (2008). Characterization of the essential activities of *Saccharomyces cerevisiae* Mtr4p, a 3'–>5' helicase partner of the nuclear exosome. *J. Biol. Chem.*, 283(8), 4930–4942.

Bini, E., Dikshit, V., Dirksen, K., Drozda, M., Blum, P. (2002). Stability of mRNA in the hyperthermophilic archaeon *Sulfolobus solfataricus*. *RNA*, 8(9), 1129–1136.

Bortolin, M.-L., Bachellerie, J.-P., Clouet-d'Orval, B. (2003). In vitro RNP assembly and methylation guide activity of an unusual box C/D RNA, cis-acting archaeal pre-tRNA(Trp). *Nucleic Acids Res.*, 31(22), 6524–6535.

Breuer, R., Gomes-Filho, J.-V., Randau, L. (2021). Conservation of archaeal C/D box sRNA-guided RNA modifications. *Front Microbiol.*, 12, 654029.

Briani, F., Carzaniga, T., Dehò, G. (2016). Regulation and functions of bacterial PNPase. *Wiley Interdiscip. Rev. RNA*, 7(2), 241–258.

Bubeck, D., Reijns, M.A.M., Graham, S.C., Astell, K.R., Jones, E.Y., Jackson, A.P. (2011). PCNA directs type 2 RNase H activity on DNA replication and repair substrates. *Nucleic Acids Res.*, 39(9), 3652–3666.

Buckley, R.J., Kramm, K., Cooper, C.D.O., Grohmann, D., Bolt, E.L. (2020). Mechanistic insights into Lhr helicase function in DNA repair. *Biochem. J.*, 477(16), 2935–2947.

Büttner, K., Wenig, K., Hopfner, K.-P. (2005). Structural framework for the mechanism of archaeal exosomes in RNA processing. *Mol. Cell.*, 20(3), 461–471.

Büttner, K., Nehring, S., Hopfner, K.-P. (2007). Structural basis for DNA duplex separation by a superfamily-2 helicase. *Nat. Struct. Mol. Biol.*, 14(7), 647–652.

Calvin, K. and Li, H. (2008). RNA-splicing endonuclease structure and function. *Cell. Mol. Life Sci.*, 65(7/8), 1176–1185.

Cao, J., Birien, T., Gayet, N., Huang, Z., Shao, Z., Jebbar, M., Alain, K. (2017). Desulfuro-bacterium indicum sp. nov., a thermophilic sulfur-reducing bacterium from the Indian Ocean. *Int. J. Syst. Evol. Microbiol.*, 67(6), 1665–1668.

Chamieh, H., Ibrahim, H., Kozah, J. (2016). Genome-wide identification of SF1 and SF2 helicases from archaea. *Gene.*, 576(1/2), 214–228.

Chan, P.P. and Lowe, T.M. (2009). GtRNAdb: A database of transfer RNA genes detected in genomic sequence. *Nucleic Acids Res.*, 37, D93–D97.

Chan, P.P., Cozen, A.E., Lowe, T.M. (2011). Discovery of permuted and recently split transfer RNAs in Archaea. *Genome Biol.*, 12(4), R38.

Charpentier, B., Fourmann, J., Branlant, C. (2007). Reconstitution of Archaeal H/ACA sRNPs and test of their activity. *Methods Enzymol.*, 2007(425), 389–405. doi: 10.1016/S0076-6879(07)25016-5.

Cho, I.-M., Lai, L.B., Susanti, D., Mukhopadhyay, B., Gopalan, V. (2010). Ribosomal protein L7Ae is a subunit of archaeal RNase P. *Proc. Natl. Acad. Sci. USA*, 107(33), 14573–14578.

Clouet-d'Orval, B., Rinaldi, D., Quentin, Y., Carpousis, A.J. (2010). Euryarchaeal beta-CASP proteins with homology to bacterial RNase J Have 5'- to 3'-exoribonuclease activity. *J. Biol. Chem.*, 285(23), 17574–17583.

Clouet-d'Orval, B., Phung, D.K., Langendijk-Genevaux, P.S., Quentin, Y. (2015). Universal RNA-degrading enzymes in Archaea: Prevalence, activities and functions of β-CASP ribonucleases. *Biochimie*, 118, 278–285.

Clouet-d'Orval, B., Batista, M., Bouvier, M., Quentin, Y., Fichant, G., Marchfelder, A., Maier, M.-K. (2018). Insights into RNA-processing pathways and associated RNA-degrading enzymes in Archaea. *FEMS Microbiol. Rev.*, 42(5), 579–613.

Condon, C. and Putzer, H. (2002). The phylogenetic distribution of bacterial ribonucleases. *Nucleic Acids Res.*, 30(24), 5339–5346.

De Falco, M., Massa, F., Rossi, M., De Felice, M. (2018). The Sulfolobus solfataricus RecQ-like DNA helicase Hel112 inhibits the NurA/HerA complex exonuclease activity. *Extremophiles*, 22(4), 581–589.

De Felice, M., Aria, V., Esposito, L., De Falco, M., Pucci, B., Rossi, M., Pisani, F.M. (2007). A novel DNA helicase with strand-annealing activity from the crenarchaeon *Sulfolobus solfataricus*. *Biochem. J.*, 408(1), 87–95.

Dennis, P.P., Tripp, V., Lui, L., Lowe, T., Randau, L. (2015). C/D box sRNA-guided 2'-O-methylation patterns of archaeal rRNA molecules. *BMC Genomics*, 16, 632.

Dominski, Z., Carpousis, A.J., Clouet-d'Orval, B. (2013). Emergence of the β-CASP ribonucleases: Highly conserved and ubiquitous metallo-enzymes involved in messenger RNA maturation and degradation. *Biochim. Biophys. Acta.*, 1829(6/7), 532–551.

Dorléans, A., Li de la Sierra-Gallay, I., Piton, J., Zig, L., Gilet, L., Putzer, H., Condon, C. (2011). Molecular basis for the recognition and cleavage of RNA by the bifunctional 5'–3' exo/endoribonuclease RNase J. *Structure*, 19(9), 1252–1261.

Durand, S. and Condon, C. (2018). RNases and helicases in Gram-positive bacteria. *Microbiol. Spectr.*, 6(2). doi: 10.1128/microbiolspec.RWR-0003-2017.

Ejaz, A. and Shuman, S. (2018). Characterization of Lhr–Core DNA helicase and manganese-dependent DNA nuclease components of a bacterial gene cluster encoding nucleic acid repair enzymes. *J. Biol. Chem.*, 293(45), 17491–17504.

Esakova, O. and Krasilnikov, A.S. (2010). Of proteins and RNA: The RNase P/MRP family. *RNA*, 16(9), 1725–1747.

Evans, D., Marquez, S.M., Pace, N.R. (2006). RNase P: Interface of the RNA and protein worlds. *Trends. Biochem. Sci.*, 31(6), 333–341.

Even, S., Pellegrini, O., Zig, L., Labas, V., Vinh, J., Bréchemmier-Baey, D., Putzer, H. (2005). Ribonucleases J1 and J2: Two novel endoribonucleases in *B. subtilis* with functional homology to *E. coli* RNase E. *Nucleic Acids Res.*, 33(7), 2141–2152.

Evguenieva-Hackenberg, E. and Klug, G. (2009). RNA degradation in Archaea and Gram-negative bacteria different from *Escherichia coli*. *Prog. Mol. Biol. Transl. Sci.*, 85, 275–317.

Evguenieva-Hackenberg, E. and Klug, G. (2011). New aspects of RNA processing in prokaryotes. *Curr. Opin. Microbiol.*, 14(5), 587–592.

Evguenieva-Hackenberg, E., Walter, P., Hochleitner, E., Lottspeich, F., Klug, G. (2003). An exosome-like complex in *Sulfolobus solfataricus*. *EMBO Rep.*, 4(9), 889–893.

Evguenieva-Hackenberg, E., Roppelt, V., Finsterseifer, P., Klug, G. (2008). Rrp4 and Csl4 are needed for efficient degradation but not for polyadenylation of synthetic and natural RNA by the archaeal exosome. *Biochemistry*, 47(50), 13158–13168.

Evguenieva-Hackenberg, E., Hou, L., Glaeser, S., Klug, G. (2014). Structure and function of the archaeal exosome. *Wiley Interdiscip. Rev. RNA*, 5(5), 623–635.

Evguenieva-Hackenberg, E., Gauernack, S., Klug, G. (2017). The archaeal exosome: Degradation and tailing at the 3′-end of RNA. In *RNA Metabolism and Gene Expression in Archaea*, Clouet-d'Orval, B. (ed.) Springer, New York.

Fairman-Williams, M.E., Guenther, U.-P., Jankowsky, E. (2010). SF1 and SF2 helicases: Family matters. *Curr. Opin. Struct. Biol.*, 20(3), 313–324.

Fischer, S., Benz, J., Späth, B., Jellen-Ritter, A., Heyer, R., Dörr, M., Maier, L.-K., Menzel-Hobeck, C., Lehr, M., Jantzer, K. et al. (2011). Regulatory RNAs in *Haloferax volcanii*. *Biochem. Soc. Trans.*, 39(1), 159–162.

Frank, D.N. and Pace, N.R. (1998). Ribonuclease P: Unity and diversity in a tRNA processing ribozyme. *Annu. Rev. Biochem.*, 67, 153–180.

Fujishima, K., Sugahara, J., Miller, C.S., Baker, B.J., Di Giulio, M., Takesue, K., Sato, A., Tomita, M., Banfield, J.F., Kanai, A. (2011). A novel three-unit tRNA splicing endonuclease found in ultrasmall Archaea possesses broad substrate specificity. *Nucleic Acids Res.*, 39(22), 9695–3704.

Gaspin, C., Cavaillé, J., Erauso, G., Bachellerie, J.P. (2000). Archaeal homologs of eukaryotic methylation guide small nucleolar RNAs: Lessons from the Pyrococcus genomes. *J. Mol. Biol.*, 297(4), 895–906.

Gauernack, A.S., Lassek, C., Hou, L., Dzieciolowski, J., Evguenieva-Hackenberg, E., Klug, G. (2017). Nop5 interacts with the archaeal RNA exosome. *FEBS Lett.*, 591(24), 4039–4048.

Gopalan, V. (2007). Uniformity amid diversity in RNase P. *Proc. Natl. Acad. Sci. USA*, 104(7), 2031–2032.

Gupta, A. and Swati, D. (2019). Riboswitches in Archaea. *Comb. Chem. High Throughput Screen*, 22(2), 135–149.

Guy, C.P. and Bolt, E.L. (2005). Archaeal Hel308 helicase targets replication forks in vivo and in vitro and unwinds lagging strands. *Nucleic Acids Res.*, 33(11), 3678–3690.

Hajj, M., Langendijk-Genevaux, P., Batista, M., Quentin, Y., Laurent, S., Capeyrou, R., Abdel-Razzak, Z., Flament, D., Chamieh, H., Fichant, G. (2021). Phylogenetic diversity of Lhr proteins and biochemical activities of the Thermococcales aLhr2 DNA/RNA helicase. *Biomolecules*, 11(7), 950. doi: 10.3390/biom11070950.

Halbach, F., Rode, M., Conti, E. (2012). The crystal structure of *S. cerevisiae* Ski2, a DExH helicase associated with the cytoplasmic functions of the exosome. *RNA*, 18(1), 124–134.

Halbach, F., Reichelt, P., Rode, M., Conti, E. (2013). The yeast ski complex: Crystal structure and RNA channeling to the exosome complex. *Cell*, 154(4), 814–826.

Hale, C., Kleppe, K., Terns, R.M., Terns, M.P. (2008). Prokaryotic silencing (psi)RNAs in *Pyrococcus furiosus*. *RNA*, 14(12), 2572–2579.

Hale, C.R., Zhao, P., Olson, S., Duff, M.O., Graveley, B.R., Wells, L., Terns, R.M., Terns, M.P. (2009). RNA-guided RNA cleavage by a CRISPR RNA-Cas protein complex. *Cell*, 139(5), 945–956.

Hamma, T. and Ferré-D'Amaré, A.R. (2004). Structure of protein L7Ae bound to a K-turn derived from an archaeal box H/ACA sRNA at 1.8 A resolution. *Structure*, 12(5), 893–903.

Hasenöhrl, D., Lombo, T., Kaberdin, V., Londei, P., Bläsi, U. (2008). Translation initiation factor a/eIF2(-gamma) counteracts 5' to 3' mRNA decay in the archaeon *Sulfolobus solfataricus*. *Proc. Natl. Acad. Sci. USA*, 105(6), 2146–2150.

Hennigan, A.N. and Reeve, J.N. (1994). mRNAs in the methanogenic archaeon *Methanococcus vannielii*: Numbers, half-lives and processing. *Mol. Microbiol.*, 11(4), 655–670.

Hirata, A., Fujishima, K., Yamagami, R., Kawamura, T., Banfield, J.F., Kanai, A., Hori, H (2012). X-ray structure of the fourth type of archaeal tRNA splicing endonuclease: Insights into the evolution of a novel three-unit composition and a unique loop involved in broad substrate specificity. *Nucleic Acids Res.*, 40(20), 10554–10566.

Hochstrasser, M.L. and Doudna, J.A. (2015). Cutting it close: CRISPR-associated endoribonuclease structure and function. *Trends Biochem. Sci.*, 40(1), 58–66.

Hölzle, A., Fischer, S., Heyer, R., Schütz, S., Zacharias, M., Walther, P., Allers T., Marchfelder, A. (2008). Maturation of the 5S rRNA 5' end is catalyzed in vitro by the endonuclease tRNase Z in the archaeon *H. volcanii*. *RNA*, 14(5), 928–937.

Hotto, A.M., Stern, D.B., Schuster, G. (2020). Plant ribonuclease J: An essential player in maintaining chloroplast RNA quality control for gene expression. *Plants*, 9(3), 334. doi: 10.3390/plants9030334.

Hou, L., Klug, G., Evguenieva-Hackenberg, E. (2013). The archaeal DnaG protein needs Csl4 for binding to the exosome and enhances its interaction with adenine-rich RNAs. *RNA Biol.*, 10(3), 415–424.

Howard, M.J., Liu, X., Lim, W.H., Klemm, B.P., Fierke, C.A., Koutmos, M., Engelke, D.R. (2013). RNase P enzymes: Divergent scaffolds for a conserved biological reaction. *RNA Biol.*, 10(6), 909–914.

Ishii, R., Minagawa, A., Takaku, H., Takagi, M., Nashimoto, M., Yokoyama, S. (2005). Crystal structure of the tRNA 3' processing endoribonuclease tRNase Z from *Thermotoga maritima*. *J. Biol. Chem.*, 280(14), 14138–14144.

Ishii, R., Minagawa, A., Takaku, H., Takagi, M., Nashimoto, M., Yokoyama, S. (2007). The structure of the flexible arm of *Thermotoga maritima* tRNase Z differs from those of homologous enzymes. *Acta. Crystallogr. Sect. F. Struct. Biol. Cryst. Commun.*, 63(8), 637–641.

Jäger, A., Samorski, R., Pfeifer, F., Klug, G. (2002). Individual gvp transcript segments in *Haloferax mediterranei* exhibit varying half-lives, which are differentially affected by salt concentration and growth phase. *Nucleic Acids Res.*, 30(24), 5436–5443.

Jäger, D., Sharma, C.M., Thomsen, J., Ehlers, C., Vogel, J., Schmitz, R.A. (2009). Deep sequencing analysis of the *Methanosarcina mazei* Gö1 transcriptome in response to nitrogen availability. *Proc. Natl. Acad. Sci. USA*, 106(51), 21878–21882.

Jäger, D., Förstner, K.U., Sharma, C.M., Santangelo, T.J., Reeve, J.N. (2014). Primary transcriptome map of the hyperthermophilic archaeon *Thermococcus kodakarensis*. *BMC Genomics*, 15, 684.

Jankowsky, E. and Fairman, M.E. (2007). RNA helicases – One fold for many functions. *Curr. Opin. Struct. Biol.*, 17(3), 316–324.

Jarrous, N. and Gopalan, V. (2010). Archaeal/eukaryal RNase P: Subunits, functions and RNA diversification. *Nucleic Acids Res.*, 38(22), 7885–7894.

Jaschinski, K., Babski, J., Lehr, M., Burmester, A., Benz, J., Heyer, R., Dörr, M., Marchfelder, A., Soppa, J. (2014). Generation and phenotyping of a collection of sRNA gene deletion mutants of the haloarchaeon *Haloferax volcanii*. *PLoS ONE*, 9(3), e90763.

Johnson, S.J. and Jackson, R.N. (2013). Ski2-like RNA helicase structures: Common themes and complex assemblies. *RNA Biol.*, 10(1), 33–43.

Khemici, V. and Linder, P. (2018). RNA helicases in RNA decay. *Biochem. Soc. Trans.*, 46(1), 163–172.

Kimura, M. and Kakuta, Y. (2012). Structural biology of the ribonuclease P in the hyperthermophilic Archaeon *Pyrococcus horikoshii* OT3. In *Microorganisms in Sustainable Agriculture and Biotechnology*, Prakash, A., Satyanarayana, T., Johri, B.N. (eds). Springer, Dordrecht.

Kitamura, S., Fujishima, K., Sato, A., Tsuchiya, D., Tomita, M., Kanai, A. (2010). Characterization of RNase HII substrate recognition using RNase HII-argonaute chimaeric enzymes from *Pyrococcus furiosus*. *Biochem. J.*, 426(3), 337–344.

Klemm, B.P., Wu, N., Chen, Y., Liu, X., Kaitany, K.J., Howard, M.J., Fierke, C.A. (2016). The diversity of ribonuclease P: Protein and RNA catalysts with analogous biological functions. *Biomolecules*, 6(2), 27. doi: 10.3390/biom6020027.

Kochiwa, H., Tomita, M., Kanai, A. (2007). Evolution of ribonuclease H genes in prokaryotes to avoid inheritance of redundant genes. *BMC Evol. Biol.*, 7, 128.

Kögel, A., Keidel, A., Bonneau, F., Schäfer, I.B., Conti, E. (2022). The human SKI complex regulates channeling of ribosome-bound RNA to the exosome via an intrinsic gatekeeping mechanism. *Mol. Cell.*, 82(4), 756–769.e8.

Koonin, E.V., Makarova, K.S., Wolf, Y.I. (2017). Evolutionary genomics of defense systems in archaea and bacteria. *Annu. Rev. Microbiol.*, 71, 233–261.

Kouzuma, Y., Mizoguchi, M., Takagi, H., Fukuhara, H., Tsukamoto, M., Numata, T., Kimura, M. (2003). Reconstitution of archaeal ribonuclease P from RNA and four protein components. *Biochem. Biophys. Res. Commun.*, 306(3), 666–673.

Kuhn, J.F., Tran, E.J., Maxwell, E.S. (2002). Archaeal ribosomal protein L7 is a functional homolog of the eukaryotic 15.5kD/Snu13p snoRNP core protein. *Nucleic Acids Res.*, 30(4), 931–941.

Lai, L., Yokota, H., Hung, L.W., Kim, R., Kim, S.H. (2000). Crystal structure of archaeal RNase HII: A homologue of human major RNase H. *Structure*, 18(8), 897–904.

Lai, L.B., Vioque, A., Kirsebom, L.A., Gopalan, V. (2010). Unexpected diversity of RNase P, an ancient tRNA processing enzyme: Challenges and prospects. *FEBS Lett.*, 584(2), 287–296.

Lai, S.M., Lai, L.B., Foster, M.P., Gopalan, V. (2014). The L7Ae protein binds to two kinkturns in the *Pyrococcus furiosus* RNase P RNA. *Nucleic Acids Res.*, 42(21), 13328–13338.

Lebreton, A. and Séraphin, B. (2008). Exosome-mediated quality control: Substrate recruitment and molecular activity. *Biochim. Biophys. Acta.*, 1779(9), 558–565.

Levy, S., Portnoy, V., Admon, J., Schuster, G. (2011). Distinct activities of several RNase J proteins in methanogenic archaea. *RNA Biol.*, 8(6), 1073–1083.

Li, H. and Abelson, J. (2000). Crystal structure of a dimeric archaeal splicing endonuclease. *J. Mol. Biol.*, 302(3), 639–648.

Li, L. and Ye, K. (2006). Crystal structure of an H/ACA box ribonucleoprotein particle. *Nature*, 443(7109), 302–307.

Li, H., Trotta, C.R., Abelson, J. (1998). Crystal structure and evolution of a transfer RNA splicing enzyme. *Science*, 280(5361), 279–284.

Li, J., Qi, L., Guo, Y., Yue, L., Li, Y., Ge, W., Wu, J., Shi, W., Dong, X. (2015). Global mapping transcriptional start sites revealed both transcriptional and post-transcriptional regulation of cold adaptation in the methanogenic archaeon *Methanolobus psychrophilus*. *Sci. Rep.*, 5, 9209.

Li de la Sierra-Gallay, I., Mathy, N., Pellegrini, O., Condon, C. (2006). Structure of the ubiquitous 3' processing enzyme RNase Z bound to transfer RNA. *Nat. Struct. Mol. Biol.*, 13(4), 376–377.

Li de la Sierra-Gallay, I., Zig, L., Jamalli, A., Putzer, H. (2008). Structural insights into the dual activity of RNase. *J. Nat. Struct. Mol. Biol.*, 15(2), 206–212.

Liang, X.-H., Liu, Q., Fournier, M.J. (2009). Loss of rRNA modifications in the decoding center of the ribosome impairs translation and strongly delays pre-rRNA processing. *RNA*, 15(9), 1716–1728.

Lim, J., Thomas, T., Cavicchioli, R. (2000). Low temperature regulated DEAD-box RNA helicase from the Antarctic archaeon, *Methanococcoides burtonii*. *J. Mol. Biol.*, 297(3), 553–567.

Lin-Chao, S., Chiou, N.-T., Schuster, G. (2007). The PNPase, exosome and RNA helicases as the building components of evolutionarily-conserved RNA degradation machines. *J. Biomed. Sci.*, 14(4), 523–532.

Linder, P. and Jankowsky, E. (2011). From unwinding to clamping – The DEAD box RNA helicase family. *Nat. Rev. Mol. Cell. Biol.*, 12(8), 505–516.

Linder, P., Lemeille, S., Redder, P. (2014). Transcriptome-wide analyses of 5'-ends in RNase J mutants of a Gram-positive pathogen reveal a role in RNA maturation, regulation and degradation. *PLoS Genet.*, 10(2), e1004207.

Londei, P. (2005). Evolution of translational initiation: New insights from the archaea. *FEMS Microbiol. Rev.*, 29(2), 185–200.

Londei, P. and Ferreira-Cerca, S. (2021). Ribosome biogenesis in archaea. *Front Microbiol.*, 12, 686977.

Lopes, R.R.S., Kessler, A.C., Polycarpo, C., Alfonzo, J.D. (2015). Cutting, dicing, healing and sealing: The molecular surgery of tRNA. *Wiley Interdiscip. Rev. RNA*, 6(3), 337–349.

Lorentzen, E., Walter, P., Fribourg, S., Evguenieva-Hackenberg, E., Klug, G., Conti, E. (2005). The archaeal exosome core is a hexameric ring structure with three catalytic subunits. *Nat. Struct. Mol. Biol.*, 12(7), 575–581.

Lorentzen, E., Dziembowski, A., Lindner, D., Seraphin, B., Conti, E. (2007). RNA channelling by the archaeal exosome. *EMBO Rep.*, 8(5), 470–476.

Lykke-Andersen, J. and Garrett, R.A. (1997). RNA-protein interactions of an archaeal homo-tetrameric splicing endoribonuclease with an exceptional evolutionary history. *EMBO J.*, 16(20), 6290–6300.

Maier, L.-K. and Marchfelder, A. (2019). It's all about the T: Transcription termination in archaea. *Biochem. Soc. Trans.*, 47(1), 461–468.

Makarova, K.S., Wolf, Y.I., Alkhnbashi, O.S., Costa, F., Shah, S.A., Saunders, S.J., Barrangou, R., Brouns, S.J.J., Charpentier, E., Haft, D.H. et al. (2015). An updated evolutionary classification of CRISPR-Cas systems. *Nat. Rev. Microbiol.*, 13(11), 722–736.

Makarova, K.S., Wolf, Y.I., Iranzo, J., Shmakov, S.A., Alkhnbashi, O.S., Brouns, S.J.J., Charpentier, E., Cheng, D., Haft, D.H., Horvath, P. et al. (2020). Evolutionary classification of CRISPR-Cas systems: A burst of class 2 and derived variants. *Nat. Rev. Microbiol.*, 18(2), 67–83.

Mao, H., White, S.A., Williamson, J.R. (1999). A novel loop-loop recognition motif in the yeast ribosomal protein L30 autoregulatory RNA complex. *Nat. Struct. Biol.*, 6(12), 1139–1147.

Marck, C. and Grosjean, H. (2003). Identification of BHB splicing motifs in intron-containing tRNAs from 18 archaea: Evolutionary implications. *RNA*, 9(12), 1516–1531.

Marini, D.B., Zhang, Y.P., Rowhani, A., Uyemoto, J.K. (2002). Etiology and host range of a closterovirus associated with plum bark necrosis-stem pitting disease. *Plant Dis.*, 86(4), 415–417.

Märtens, B., Amman, F., Manoharadas, S., Zeichen, L., Orell, A., Albers, S.-V., Hofacker, I., Bläsi, U. (2013). Alterations of the transcriptome of *Sulfolobus acidocaldarius* by exoribonuclease aCPSF2. *PLoS ONE*, 8(10), e76569.

Märtens, B., Bezerra, G.A., Kreuter, M.J., Grishkovskaya, I., Manica, A., Arkhipova, V., Djinovic-Carugo, K., Bläsi, U. (2015). The heptameric SmAP1 and SmAP2 proteins of the crenarchaeon *sulfolobus solfataricus* bind to common and distinct RNA targets. *Life (Basel)*, 5(2), 1264–1281.

Märtens, B., Hou, L., Amman, F., Wolfinger, M.T., Evguenieva-Hackenberg, E., Bläsi, U. (2017). The SmAP1/2 proteins of the crenarchaeon *Sulfolobus solfataricus* interact with the exosome and stimulate A-rich tailing of transcripts. *Nucleic Acids Res.*, 45(13), 7938–7949.

Mathy, N., Bénard, L., Pellegrini, O., Daou, R., Wen, T., Condon, C. (2007). 5'-to-3' exoribonuclease activity in bacteria: Role of RNase J1 in rRNA maturation and 5′ stability of mRNA. *Cell*, 129(4), 681–692.

Matos, R.G., López-Viñas, E., Goméz-Puertas, P., Arraiano, C.M. (2012). The only exoribonuclease present in *Haloferax volcanii* has a unique response to temperature changes. *Biochim. Biophys. Acta.*, 1820(10), 1543–1552.

Matsuo, Y., Ikeuchi, K., Saeki, Y., Iwasaki, S., Schmidt, C., Udagawa, T., Sato, F., Tsuchiya, H., Becker, T., Tanaka, K. et al. (2017). Ubiquitination of stalled ribosome triggers ribosome-associated quality control. *Nat. Commun.*, 8(1), 159.

Meola, N., Domanski, M., Karadoulama, E., Chen, Y., Gentil, C., Pultz, D., Vitting-Seerup, K., Lykke-Andersen, S., Andersen, J.S., Sandelin, A., Jensen, T.H. (2016). Identification of a nuclear exosome decay pathway for processed transcripts. *Mol. Cell.*, 64(3), 520–533.

Mir-Montazeri, B., Ammelburg, M., Forouzan, D., Lupas, A.N., Hartmann, M.D. (2011). Crystal structure of a dimeric archaeal cleavage and polyadenylation specificity factor. *J. Struct. Biol.*, 173(1), 191–195.

Mura, C., Randolph, P.S., Patterson, J., Cozen, A.E. (2013). Archaeal and eukaryotic homologs of Hfq: A structural and evolutionary perspective on Sm function. *RNA Biol.*, 10(4), 636–651.

Muroya, A., Tsuchiya, D., Ishikawa, M., Haruki, M., Morikawa, M., Kanaya, S., Morikawa, K. (2001). Catalytic center of an archaeal type 2 ribonuclease H as revealed by X-ray crystallographic and mutational analyses. *Protein Sci.*, 10(4), 707–714.

Newman, J.A., Hewitt, L., Rodrigues, C., Solovyova, A., Harwood, C.R., Lewis, R.J. (2011). Unusual, dual endo- and exonuclease activity in the degradosome explained by crystal structure analysis of RNase J1. *Structure*, 19(9), 1241–1251.

Nishida, Y., Ishikawa, H., Baba, S., Nakagawa, N., Kuramitsu, S., Masui, R. (2010). Crystal structure of an archaeal cleavage and polyadenylation specificity factor subunit from Pyrococcus horikoshii. *Proteins*, 78(10), 2395–2398.

Nolivos, S., Carpousis, A.J., Clouet-d'Orval, B. (2005). The K-loop, a general feature of the Pyrococcus C/D guide RNAs, is an RNA structural motif related to the K-turn. *Nucleic Acids Res.*, 33(20), 6507–6514.

Omer, A.D., Ziesche, S., Ebhardt, H., Dennis, P.P. (2002). In vitro reconstitution and activity of a C/D box methylation guide ribonucleoprotein complex. *Proc. Natl. Acad. Sci. USA*, 99(8), 5289–5294.

Patrick, E.M., Srinivasan, S., Jankowsky, E., Comstock, M.J. (2017). The RNA helicase Mtr4p is a duplex-sensing translocase. *Nat. Chem. Biol.*, 13(1), 99–104.

Peterson, J.R., Thor, S., Kohler, L., Kohler, P.R.A., Metcalf, W.W., Luthey-Schulten, Z. (2016). Genome-wide gene expression and RNA half-life measurements allow predictions of regulation and metabolic behavior in Methanosarcina acetivorans. *BMC Genomics*, 17(1), 924.

Phok, K., Moisan, A., Rinaldi, D., Brucato, N., Carpousis, A.J., Gaspin, C., Clouet-d'Orval, B. (2011). Identification of CRISPR and riboswitch related RNAs among novel noncoding RNAs of the euryarchaeon Pyrococcus abyssi. *BMC Genomics*, 12(312).

Phung, D.K., Rinaldi, D., Langendijk-Genevaux, P.S., Quentin, Y., Carpousis, A.J., Clouet-d'Orval, B. (2013). Archaeal β-CASP ribonucleases of the aCPSF1 family are orthologs of the eukaryal CPSF-73 factor. *Nucleic Acids Res.*, 41(2), 1091–1103.

Phung, D.K., Etienne, C., Batista, M., Langendijk-Genevaux, P., Moalic, Y., Laurent, S., Liuu, S., Morales, V., Jebbar, M., Fichant, G. et al. (2020). RNA processing machineries in Archaea: The 5'–3' exoribonuclease aRNase J of the β-CASP family is engaged specifically with the helicase ASH-Ski2 and the 3'–5' exoribonucleolytic RNA exosome machinery. *Nucleic Acids Res.*, 48(7), 3832–3847.

Pluchon, P.-F., Fouqueau, T., Crezé, C., Laurent, S., Briffotaux, J., Hogrel, G., Palud, A., Henneke, G., Godfroy, A., Hausner, W. et al. (2013). An extended network of genomic maintenance in the archaeon *Pyrococcus abyssi* highlights unexpected associations between eucaryotic homologs. *PLoS ONE*, 8(11), e79707.

Portnoy, V. and Schuster, G. (2006). RNA polyadenylation and degradation in different Archaea. Roles of the exosome and RNase R. *Nucleic Acids Res.*, 34(20), 5923–5931.

Qi, L., Li, J., Jia, J., Yue, L., Dong, X. (2020). Comprehensive analysis of the pre-ribosomal RNA maturation pathway in a methanoarchaeon exposes the conserved circularization and linearization mode in archaea. *RNA Biol.*, 17(10), 1427–1441.

Ramos, C.R.R., Oliveira, C.L.P., Torriani, I.L., Oliveira, C.C. (2006). The Pyrococcus exosome complex: Structural and functional characterization. *J. Biol. Chem.*, 281(10), 6751–6759.

Randau, L., Calvin, K., Hall, M., Yuan, J., Podar, M., Li, H., Söll, D. (2005). The heteromeric *Nanoarchaeum equitans* splicing endonuclease cleaves noncanonical bulge-helix-bulge motifs of joined tRNA halves. *Proc. Natl. Acad. Sci. USA*, 102(50), 17934–1739.

Redder, P., Hausmann, S., Khemici, V., Yasrebi, H., Linder, P. (2015). Bacterial versatility requires DEAD-box RNA helicases. *FEMS Microbiol. Rev.*, 39(3), 392–412.

Redko, Y., Li de la Sierra-Gallay, I., Condon, C. (2007). When all's zed and done: The structure and function of RNase Z in prokaryotes. *Nat. Rev. Microbiol.*, 5(4), 278–286.

Rinke, C., Schwientek, P., Sczyrba, A., Ivanova, N.N., Anderson, I.J., Cheng, J.-F., Darling, A., Malfatti, S., Swan, B.K., Gies, E.A. et al. (2013). Insights into the phylogeny and coding potential of microbial dark matter. *Nature*, 499(7459), 431–437.

Roppelt, V., Klug, G., Evguenieva-Hackenberg, E. (2010). The evolutionarily conserved subunits Rrp4 and Csl4 confer different substrate specificities to the archaeal exosome. *FEBS Lett.*, 584(13), 2931–2936.

Rozhdestvensky, T.S., Tang, T.H., Tchirkova, I.V., Brosius, J., Bachellerie, J.-P., Hüttenhofer, A. (2003). Binding of L7Ae protein to the K-turn of archaeal snoRNAs: A shared RNA binding motif for C/D and H/ACA box snoRNAs in Archaea. *Nucleic Acids Res.*, 31(3), 869–877.

Samanta, M.P., Lai, S.M., Daniels, C.J., Gopalan, V. (2016). Sequence analysis and comparative study of the protein subunits of archaeal rnase P. *Biomolecules*, 6(2), 22. doi: 10.3390/biom6020022.

Sanders, T.J., Wenck, B.R., Selan, J.N., Barker, M.P., Trimmer, S.A., Walker, J.E., Santangelo, T.J. (2020). FttA is a CPSF73 homologue that terminates transcription in Archaea. *Nat. Microbiol.*, 5(4), 545–553.

Schaeffer, D., Tsanova, B., Barbas, A., Reis, F.P., Dastidar, E.G., Sanchez-Rotunno, M., Arraiano, C.M., van Hoof, A. (2009). The exosome contains domains with specific endoribonuclease, exoribonuclease and cytoplasmic mRNA decay activities. *Nat. Struct. Mol. Biol.*, 16(1), 56–62.

Schierling, K., Rösch, S., Rupprecht, R., Schiffer, S., Marchfelder, A. (2002). tRNA 3' end maturation in archaea has eukaryotic features: The RNase Z from *Haloferax volcanii*. *J. Mol. Biol.*, 316(4), 895–902.

Schmidt, C., Kowalinski, E., Shanmuganathan, V., Defenouillère, Q., Braunger, K., Heuer, A., Pech, M., Namane, A., Berninghausen, O., Fromont-Racine, M. et al. (2016). The cryo-EM structure of a ribosome-Ski2-Ski3-Ski8 helicase complex. *Science*, 354(6318), 1431–1433.

Schwarz, T.S., Berkemer, S.J., Bernhart, S.H., Weiß, M., Ferreira-Cerca, S., Stadler, P.F., Marchfelder, A. (2020). Splicing endonuclease is an important player in rRNA and tRNA maturation in Archaea. *Front Microbiol.*, 11, 594838.

Sharwood, R.E., Halpert, M., Luro, S., Schuster, G., Stern, D.B. (2011). Chloroplast RNase J compensates for inefficient transcription termination by removal of antisense RNA. *RNA*, 17(12), 2165–2176.

Shimada, Y., Fukuda, W., Akada, Y., Ishida, M., Nakayama, J., Imanaka, T., Fujiwara, S. (2009). Property of cold inducible DEAD-box RNA helicase in hyperthermophilic archaea. *Biochem. Biophys. Res. Commun.*, 389(4), 622–627.

Silva, A.P., Chechik, M., Byrne, R.T., Waterman, D.G., Ng, C.L., Dodson, E.J., Koonin, E.V., Antson, A.A., Smits, C. (2011). Structure and activity of a novel archaeal beta-CASP protein with N-terminal KH domains. *Structure*, 19(5), 622–632.

Sitron, C.S., Park, J.H., Brandman, O. (2017). Asc1, Hel2, and Slh1 couple translation arrest to nascent chain degradation. *RNA*, 23(5), 798–810.

Soppa, J., Straub, J., Brenneis, M., Jellen-Ritter, A., Heyer, R., Fischer, S., Granzow, M., Voss, B., Hess, W.R., Tjaden, B., Marchfelder, A. (2009). Small RNAs of the halophilic archaeon *Haloferax volcanii*. *Biochem. Soc. Trans.*, 37(1), 133–136.

Späth, B., Schubert, S., Lieberoth, A., Settele, F., Schütz, S., Fischer, S., Marchfelder, A. (2008). Two archaeal tRNase Z enzymes: Similar but different. *Arch. Microbiol.*, 190(3), 301–308.

Stoecklin, G. and Mühlemann, O. (2013). RNA decay mechanisms: Specificity through diversity. *Biochim. Biophys. Acta.*, 1829(6/7), 487–490.

Straub, J., Brenneis, M., Jellen-Ritter, A., Heyer, R., Soppa, J., Marchfelder, A. (2009). Small RNAs in haloarchaea: Identification, differential expression and biological function. *RNA Biol.*, 6(3), 281–292.

Tadokoro, T. and Kanaya, S. (2009). Ribonuclease H: Molecular diversities, substrate binding domains, and catalytic mechanism of the prokaryotic enzymes. *FEBS J.*, 276(6), 1482–1493.

Tafel, A.A., Wu, L., McHugh, P.J. (2011). Human HEL308 localizes to damaged replication forks and unwinds lagging strand structures. *J. Biol. Chem.*, 286(18), 15832–15840.

Tang, T.-H., Bachellerie, J.-P., Rozhdestvensky, T., Bortolin, M.-L., Huber, H., Drungowski, M., Elge, T., Brosius, J., Hüttenhofer, A. (2002). Identification of 86 candidates for small non-messenger RNAs from the archaeon *Archaeoglobus fulgidus*. *Proc. Natl. Acad. Sci. USA*, 99(11), 7536–7541.

Tang, T.-H., Polacek, N., Zywicki, M., Huber, H., Brugger, K., Garrett, R., Bachellerie, J.P., Hüttenhofer, A. (2005). Identification of novel non-coding RNAs as potential antisense regulators in the archaeon *Sulfolobus solfataricus*. *Mol. Microbiol.*, 55(2), 469–481.

Thompson, L.D. and Daniels, C.J. (1988). A tRNA(Trp) intron endonuclease from *Halobacterium volcanii*. Unique substrate recognition properties. *J. Biol. Chem.*, 263(34), 17951–17959.

Thompson, L.D. and Daniels, C.J. (1990). Recognition of exon-intron boundaries by the *Halobacterium volcanii* tRNA intron endonuclease. *J. Biol. Chem.*, 265(30), 18104–18111.

Tocchini-Valentini, G.D., Fruscoloni, P., Tocchini-Valentini, G.P. (2005). Coevolution of tRNA intron motifs and tRNA endonuclease architecture in Archaea. *Proc. Natl. Acad. Sci. USA*, 102(43), 15418–15422.

Toffano-Nioche, C., Ott, A., Crozat, E., Nguyen, A.N., Zytnicki, M., Leclerc, F., Forterre, P., Bouloc, P., Gautheret, D. (2013). RNA at 92°C: The non-coding transcriptome of the hyperthermophilic archaeon *Pyrococcus abyssi*. *RNA Biol.*, 10(7), 1211–1220.

Tripp, V., Martin, R., Orell, A., Alkhnbashi, O.S., Backofen, R., Randau, L. (2017). Plasticity of archaeal C/D box sRNA biogenesis. *Mol. Microbiol.*, 103(1), 151–164.

Veith, T., Martin, R., Wurm, J.P., Weis, B.L., Duchardt-Ferner, E., Safferthal, C., Hennig, R., Mirus, O., Bohnsack, M.T., Wöhnert, J., Schleiff, E. (2012). Structural and functional analysis of the archaeal endonuclease Nob1. *Nucleic Acids Res.*, 40(7), 3259–3274.

Vester, K., Santos, K.F., Kuropka, B., Weise, C., Wahl, M.C. (2020). The inactive C-terminal cassette of the dual-cassette RNA helicase BRR2 both stimulates and inhibits the activity of the N-terminal helicase unit. *J. Biol. Chem.*, 295(7), 2097–2112.

Vestergaard, G., Garrett, R.A., Shah, S.A. (2014). CRISPR adaptive immune systems of Archaea. *RNA Biol.*, 11(2), 156–167.

Vidovic, I., Nottrott, S., Hartmuth, K., Lührmann, R., Ficner, R. (2000). Crystal structure of the spliceosomal 15.5kD protein bound to a U4 snRNA fragment. *Mol. Cell.*, 6(6), 1331–1342.

Vogel, A., Schilling, O., Späth, B., Marchfelder, A. (2005). The tRNase Z family of proteins: Physiological functions, substrate specificity and structural properties. *Biol. Chem.*, 386(12), 1253–1264.

Walker, J.E., Luyties, O., Santangelo, T.J. (2017). Factor-dependent archaeal transcription termination. *Proc. Natl. Acad. Sci. USA*, 114(33), E6767–6773.

Weick, E.-M., Puno, M.R., Januszyk, K., Zinder, J.C., DiMattia, M.A., Lima, C.D. (2018). Helicase-dependent RNA decay illuminated by a cryo-EM structure of a human nuclear RNA exosome-MTR4 complex. *Cell*, 173(7), 1663–1677.e21.

Witharana, C., Roppelt, V., Lochnit, G., Klug, G., Evguenieva-Hackenberg, E. (2012). Heterogeneous complexes of the RNA exosome in *Sulfolobus solfataricus*. *Biochimie*, 94(7), 1578–1587.

Wolin, S.L. and Maquat, L.E. (2019). Cellular RNA surveillance in health and disease. *Science*, 366(6467), 822–827.

Yip, W.S.V., Vincent, N.G., Baserga, S.J. (2013). Ribonucleoproteins in archaeal pre-rRNA processing and modification. *Archaea*, 2013, 614735. doi: 10.1155/2013/614735.

Zaremba-Niedzwiedzka, K., Caceres, E.F., Saw, J.H., Bäckström, D., Juzokaite, L., Vancaester, E., Seitz, K.W., Anantharaman, K., Starnawski, P., Kjeldsen, K.U. et al. (2017). Asgard archaea illuminate the origin of eukaryotic cellular complexity. *Nature*, 541(7637), 353–358.

Zhai, B., DuPrez, K., Han, X., Yuan, Z., Ahmad, S., Xu, C., Gu, L., Ni, J., Fan, L., Shen, Y. (2018). The archaeal ATPase PINA interacts with the helicase Hjm via its carboxyl terminal KH domain remodeling and processing replication fork and Holliday junction. *Nucleic Acids Res.*, 46(13), 6627–6641.

Zhang, J. and Ferré-D'Amaré, A.R. (2016). The tRNA elbow in structure, recognition and evolution. *Life (Basel)*, 6(1), 3. doi: 10.3390/life6010003.

Zhang, L., Li, X., Hill, R.C., Qiu, Y., Zhang, W., Hansen, K.C., Zhao, R. (2015). Brr2 plays a role in spliceosomal activation in addition to U4/U6 unwinding. *Nucleic Acids Res.*, 43(6), 3286–3297.

Zheng, X., Feng, N., Li, D., Dong, X., Li, J. (2017). New molecular insights into an archaeal RNase J reveal a conserved processive exoribonucleolysis mechanism of the RNase J family. *Mol. Microbiol.*, 106(3), 351–366.

Zinoviev, A., Ayupov, R.K., Abaeva, I.S., Hellen, C.U.T., Pestova T.V. (2020). Extraction of mRNA from stalled ribosomes by the Ski complex. *Mol. Cell.*, 77(6), 1340–1349.e6.

5

Ribosome and Transfer RNA Biogenesis

Tamara BASTA[1] and Sébastien FERREIRA-CERCA[2,3]

[1]*Institute for Integrative Biology of the Cell (I2BC), CEA, CNRS, Université Paris-Saclay, Gif-sur-Yvette, France*

[2]*Cellular Biochemistry of Microorganisms, Institute for Biochemistry, Genetics and Microbiology, Regensburg Center for Biochemistry, University of Regensburg, Germany*

[3]*BIOC, CNRS, Institut Polytechnique de Paris, Palaiseau, France*

Across the tree of life, most of the components required for protein synthesis undergo extensive maturation ensuring the acquisition of their functionality. In this section, we will have a closer look at how some of the key players of the translation process are synthesized and matured in the third domain of life, the archaea. As will be described in Chapter 7 of this volume, protein synthesis is an intricate molecular choreography that requires the coordinated action of (i) the ribosome, which performs the decoding of the genetic information and the synthesis of the peptide bond between two amino acids; (ii) the messenger RNA (mRNA), which contains the genetic information; and (iii) the aminoacyl-transfer RNA (aa-tRNA), which acts as an intermediate molecule providing the right amino acid at the right position.

This section focuses on the birth and maturation of ribosomal subunits and tRNA molecules in archaea. The ribosomal subunits and tRNA biogenesis processes ensure the production of some of the core molecular components of the translation process. Despite being a distinct domain of life, archaea are often characterized by a mixture of bacterial and eukaryotic features. Ribosome and tRNA biogenesis pathways are not an exception to this rule and as such their studies provide valuable insights into the evolution of these molecular processes and ultimately the evolution

Biology of Archaea 2,
coordinated by Béatrice CLOUET-D'ORVAL, Bruno FRANZETTI and Philippe OGER.

of life on Earth. When possible, key conceptual similarities and differences between the biogenesis pathways across the different domains of life will be highlighted.

5.1. Ribosome biogenesis

5.1.1. *Ribosome composition*

First, we will focus on the molecular ingredients establishing ribosomes. The ribosome is an intricated ribonucleoprotein complex formed by the timely ordered assembly of two kinds of structural components, the ribosomal RNAs (rRNAs) and ribosomal proteins (r-proteins). The ribosome itself can be subdivided into two different ribosomal subunits, the small ribosomal subunit (SSU) and the large ribosomal subunit (LSU) (Figure 5.1).

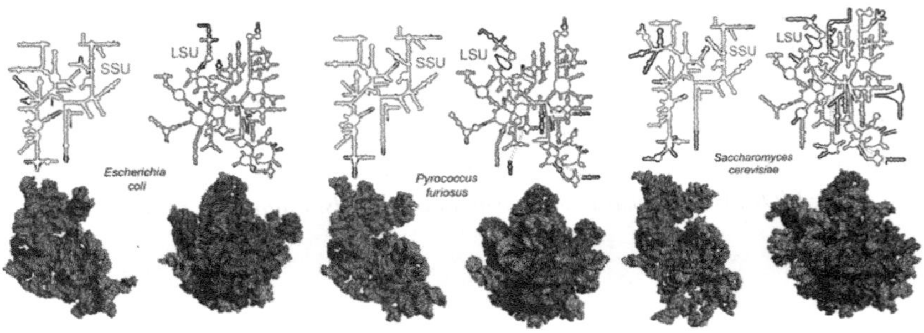

Figure 5.1. *Representative ribosomal subunit structures across the tree of life (modified from Figures 4 and 5 in Bowman et al. 2020). For a color version of this figure, see www.iste.co.uk/clouet/archaea2.zip*

COMMENT ON FIGURE 5.1.– *The universal common core of rRNA mapped onto the secondary and three-dimensional structures of rRNAs of a bacterium (lower left), an archaeon (lower middle) and a eukaryote (lower right). The SSU (red/gray) contains the 16S/18S rRNA and the LSU (blue/gray) contains the 23S/25S, the 5S and additionally the 5.8S rRNAs in eukaryotes (see also Figure 5.2). Red (SSU) and blue (LSU) indicate common core rRNA parts. Black or gray indicate rRNA that is not part of the common core and is variable in structure or absent from some species. The secondary structure of the rRNA of the bacterium E. coli (upper left), the archaeon P. furiosus (upper middle) and eukaryote S. cerevisiae (upper right) is indicated. Some sites of insertion of microexpansion or expansion segments are indicated by dashed lines in the archaeon/eukaryotes secondary structure. Each three-dimensional structure is viewed from the solvent exposed surface of the assembled ribosome, with the subunit interface directed onto the page. Ribosomal subunits structures are available at the Protein Data Bank (see: https://www.rcsb.org/) under the following accession numbers: E. coli, PDB 4V9D; P. furiosus, PDB 4V6U; S. cerevisiae, PDB 4V88.*

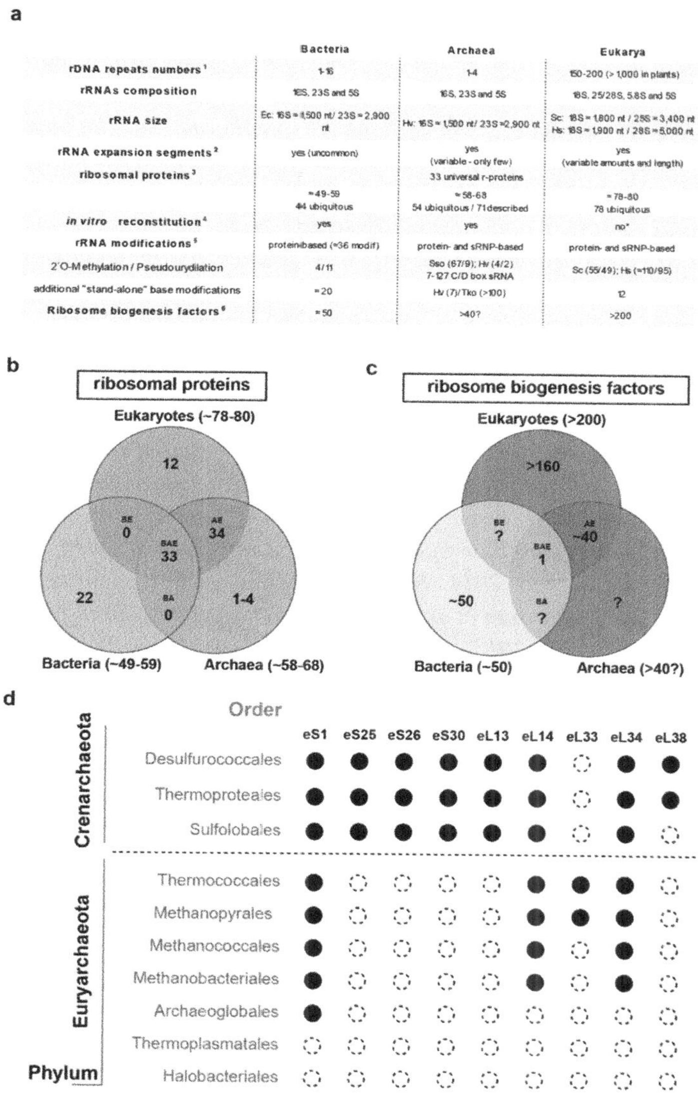

Figure 5.2. *Ribosome and ribosome biogenesis key features overview across the tree of life (reproduced from Londei and Ferreira-Cerca 2021). For a color version of this figure, see www.iste.co.uk/clouet/archaea2.zip*

COMMENT ON FIGURE 5.2.– *a) Summary of ribosome and ribosome biogenesis key features. Modified from Ferreira-Cerca (2017) according to the following previous studies: (1) Hadjiolov 1985; Warner 1999; Klappenbach et al. 2001; Stoddard et al. 2015; (2) Gerbi 1986, 1996; Armache et al. 2013; Parker et al. 2015; Petrov et al. 2015; (3) Lecompte et al. 2002; Nakao et al. 2004; Yutin et al. 2012; (4) Londei et al. 1986; Nierhaus 1991; Sanchez et al. 1996, 1990; Mangiarotti and Chiaberge 1997; Culver 2003; Nierhaus and Lafontaine 2004; (5) Lafontaine and Tollervey 1998; Grosjean et al. 2008; Dennis et al. 2015; Sharma and Lafontaine 2015; Krogh et al. 2016; Sloan et al. 2016; Taoka et al. 2018; Coureux et al. 2020; Grünberger et al. 2020; Sas-Chen et al. 2020; (6) Hage and Tollervey 2004; Thomson et al. 2013; Woolford and Baserga 2013; Ebersberger et al. 2014; Grosjean et al. 2014; Henras et al. 2015. A detailed list of putative eukaryotic ribosome biogenesis factors conserved in archaea is available in Ebersberger et al. (2014). Abbreviations used: Sso: Saccharolobus solfataricus; Hv: Haloferax volcanii; Tko: Thermococcus kodakarensis; Hs: Homo sapiens; Sc: Saccharomyces cerevisiae. b and c are a summary of shared r-proteins (b) and ribosome biogenesis factors (c) across the three domains of life. Numbers of r-proteins and putative ribosome biogenesis factors sequence homologues shared between bacteria, archaea and eukarya (BAE); bacteria, archaea (BA), archaea and eukarya (AE), bacteria and eukarya (BE), or unique to bacteria (B), archaea (A), or eukarya (E) are indicated (summary based on Ban et al. 2014; Coureux et al. 2020; Ebersberger et al. 2014; Grosjean et al. 2014; Hage and Tollervey 2004; Henras et al. 2015; Lecompte et al. 2002; Márquez et al. 2011; Nakao et al. 2004; Nürenberg-Goloub et al. 2020; Thomson et al. 2013; Woolford and Baserga 2013; Yutin et al. 2012). d shows exemplary gene distribution of selected archaeal r-proteins shared between archaea and eukaryotes across two major archaeal Phyla. The black circle denotes the presence, and open circle denotes the absence of sequence homologue for the indicated r-protein of the small (S) or large (L) ribosomal subunits, respectively. Adapted from Lecompte et al. (2002) and Yutin et al. (2012) using the nomenclature proposed in Ban et al. (2014).*

Although ribosomes are universally conserved and found in every living cell (cytosolic ribosomes) and in organelles (e.g. mitochondria), the structural component composition of the two ribosomal subunits varies significantly across and within the tree of life (Figure 5.2).

The archaeal ribosomal subunits are composed of three types of ubiquitous rRNAs, the SSU-specific 16S rRNA and the LSU-specific 23S and 5S rRNAs, to which a total of 60–70 r-proteins are associated. Among these r-proteins, 33 are universally conserved, that is, they are found in all ribosomal subunits. Remarkably, the universally conserved r-proteins and the universal prokaryotic rRNAs core establish a conserved (ancestral) ribosomal core. However, these shared molecular features show sufficient organisms-specific intrinsic primary sequence variabilities, which were originally recognized and harnessed by the pioneering studies of Carl Woese and his collaborators. These initial studies have revolutionized our thinking about the evolution of life on Earth by inferring the first molecular-based phylogenetic analysis of living organisms and by revealing the existence of a third domain of life, the archaea. Noteworthy, these molecular features still belong to the fundamental pilar of modern molecular phylogenetic analysis and taxonomy (for details, see Chapters 2 and 3, Volume 1).

Coming back to the ribosomal subunit structural components, the archaeal rRNA molecules are generally described to be of prokaryotic nature and are similar in size and composition to their bacterial counterparts. In fact, most eukaryotic cytosolic rRNAs are characterized by the presence of so-called expansion segments (ES), which are additional RNA sequences of different sizes inserted within the prokaryotic rRNA core. These insertions contribute to the noticeable rRNA size increase encountered in eukaryotic cytosolic ribosomal subunits (e.g. SSU 16S→18S or LSU 23S→25S/28S rRNAs). The size and number of ES present in the cytosolic ribosomal subunit is not homogeneously distributed and may contribute to functional diversification and/or adaptation of the eukaryotic ribosomal machinery. For example, some of these ES have been implicated in ribosome biogenesis and/or function.

In contrast to the prevalent view so far, it should be also noted that these ES are not specific idiosyncrasies of the eukaryotic rRNAs, but are also found among few bacteria and several archaea. Most of these ES are, however, limited in size and number when compared to the eukaryotic ES. Nevertheless, a few notable exceptions of larger ES, similar in size to those encountered in Eukaryotes, have been recently described in the archaeal Asgard superphylum[1].

The second type of structural components shaping the ribosomal subunits, the r-proteins, can be divided into three different groups: (1) the universally conserved r-proteins; (2) r-proteins exclusively shared between archaea and eukaryotes; and (3) archaeal specific r-proteins (Figure 5.3).

At this point, we can already highlight the absence of exclusively shared r-proteins between archaea and bacteria. Notably, the presence of shared r-proteins between archaea and eukaryotes is an important feature as it belongs to the numerous characteristics shared between archaea and eukaryotes and reflect their

1 See Figure 1 in Penev et al. (2020), available in open access at: https://academic.oup.com/gbe/article/12/10/1694/5891659.

common evolutionary relationship (see Chapter 2, Volume 1). Among the ~ 70 r-proteins described in archaea, only 54 are known to be ubiquitous based on protein sequence conservation throughout the archaeal genomes investigated so far. Consequently, the r-proteins composition also varies across archaea and largely follow the archaeal phylogenetic relationship. The third group contains the archaeal specific r-proteins. So far, this group consists of only a few r-proteins that have been recently discovered (Figure 5.3), thereby suggesting that the search for additional archaeal specific r-proteins may still be incomplete.

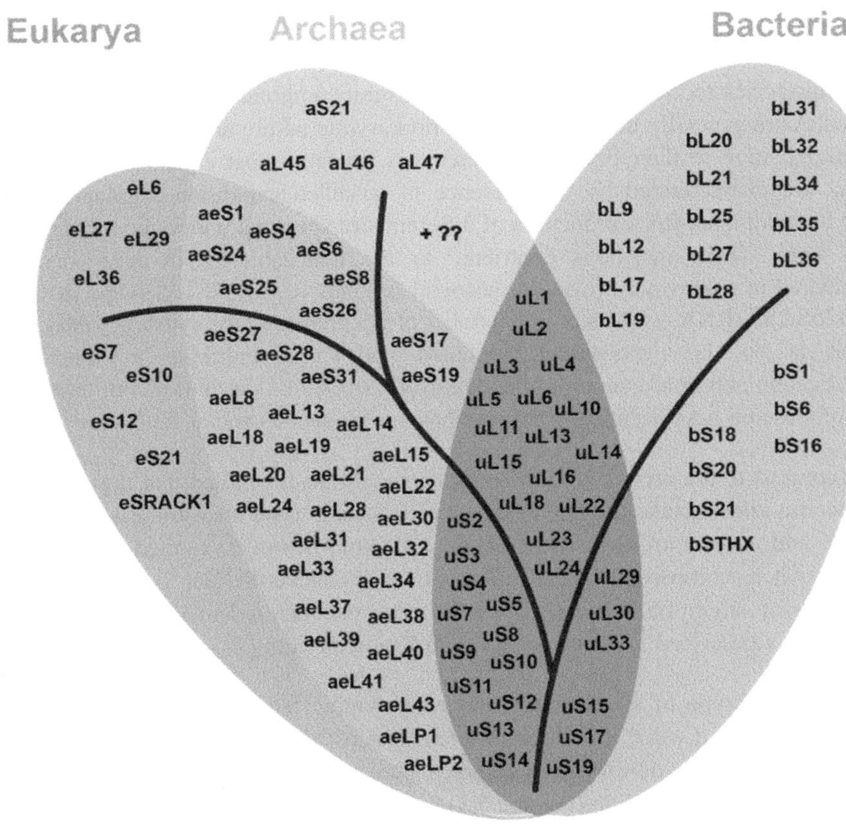

Figure 5.3. *Phylogenetic distribution of ribosomal proteins across the tree of life (adapted from Bowman et al. (2020) and Kovacs et al. (2018) using an updated and extended version of the r-proteins nomenclature proposed by Ban et al. (2014)). For a color version of this figure, see www.iste.co.uk/clouet/archaea2.zip*

COMMENT ON FIGURE 5.3.– *Venn diagram representation of the phylogenetic distribution of r-proteins across the tree of life. Universal r-proteins found in all domains of life are listed in the gray region. Bacteria-specific r-proteins are listed in the blue region on the right, the archaea-specific r-proteins in the yellow region and the eukarya-specific r-proteins in the lilac region on the left. Shared archaea–eukarya r-proteins are listed in the orange region. The extended r-protein nomenclature is based on Ban et al. (2014) and integrates information of the phylogenetic distribution consistent with the tree of life. SSU and LSU proteins are indicated as SX and LX, respectively. Phylogenetic distribution is indicated by the name prefix: u: universal; b: bacteria; a: archaea; e: eukarya; and ae: archaea–eukarya. Note that this r-protein naming scheme, by incorporating evolutionary relationships into r-protein names, is intended to facilitate understanding of the evolution of the translation machinery.*

Finally, in eukaryotic and bacterial organisms investigated thus far, the r-proteins appear in one copy per ribosome apart from the r-proteins establishing the ribosomal stalk, a structural feature of the LSU.

In contrast, some archaeal ribosomal subunits are characterized by the presence of intra- and inter-subunit promiscuous r-proteins, which, on the one hand, modifies the relative stoichiometry of these r-proteins within one ribosomal subunit and, on the other hand, show uncommon shared structural components within both SSU and LSU[2].

To summarize this part, archaeal ribosomal subunits are structurally related to their bacterial and eukaryotic relatives. Even though archaeal ribosomal subunits are more closely related to the eukaryotic ones, their compositional and underlying structural variability and peculiarities is sufficiently pronounced to distinguish them from the other ribosomal subunits found in the tree of life.

These observations already imply that significant variation between the bacterial, eukaryotic and archaeal ribosome biogenesis pathways may occur, but also potential differences within the archaeal lineage can be anticipated.

The ribosome biogenesis process will be the focus of the following sections where we will emphasize the key known steps enabling the formation of functional ribosomal subunits in archaea.

2 See Figure 6 in Armache et al. (2013), available in open access at: https://doi.org/10.1093/nar/gks1259.

5.1.2. *Ribosome biogenesis in a nutshell*

Before exploring in further details how archaeal cells assemble functional ribosomal subunits from their structural constituents – the rRNA molecules and r-proteins – it is important to define the general steps that may be logically required to achieve this critical cellular task. Once these steps have been introduced (see box below), the molecular bases underlying the molecular progression of these steps will be described individually in the following sections. When pertinent, possible crosstalk between these different molecular events will be highlighted.

According to the recipe summarized above, our first step to dive into the ribosome biogenesis pathway is to examine how synthesis of every single ribosomal constituent according to cellular needs, and necessary stoichiometry, proceeds. Importantly, some of these constituents, like the ribosomal RNAs, are produced as immature rRNA precursors (hereafter, pre-rRNAs) that are flanked by RNA sequences that will be trimmed by RNA processing activities (ribonucleases will be involved). In addition, *RNA decoration* in the form of covalent rRNA modifications will be incorporated at different stages of the process. Furthermore, the rRNAs and the r-proteins will associate together, in a timely defined and cooperative manner, to ensure proper assembly of the respective ribosomal subunits. All of these steps will be supported by additional transiently acting helpers, the ribosome biogenesis factors, also known as assembly or *trans*-acting factors (see Toolbox 5.1).

The ribosomal subunit biogenesis recipe

Making a ribosome can be compared to a cooking recipe where all of the ingredients need to be available in sufficient amounts. Some of these may be further processed prior usage and/or mixed in a particular order and/or cannot be omitted at all. Moreover, mixing or preparing these ingredients may require some additional help to efficiently mix the ribosomal subunit dough. Not forgetting that the ribosomal mixture maybe prepared in a timely controlled manner to allow for its proper maturation. To some extent, additional accessory decorations may be incorporated to improve the overall result. Finally, the product of all of these molecular efforts should be able to withstand sufficient critical quality control to optimally fulfill its final function.

Interestingly, the main outline of the recipe described above is ubiquitously valid for any ribosomal subunit biogenesis pathway. Only in its detail, this general recipe will vary across different biological systems.

Now going beyond this simplified but tangible analogy of the ribosome biogenesis process, we need to define all of the key players and steps that are ultimately necessary to synthesize archaeal ribosomes.

5.1.3. *Synthesis of the ribosomal constituents*

In archaea, the rDNA genes are found in one to four copies per haploid genome. First, we will focus on the rather peculiar but conserved rRNA gene organization. Across the tree of life, most rDNA genes are organized as polycistronic transcriptional unit that contains the mature rRNAs separated by spacer sequence elements. Consequently, the future mature rRNAs are embedded within a primary transcript containing RNA spacer elements that must be processed by endo- and/or exo-nucleolytic activities to liberate the mature rRNAs.

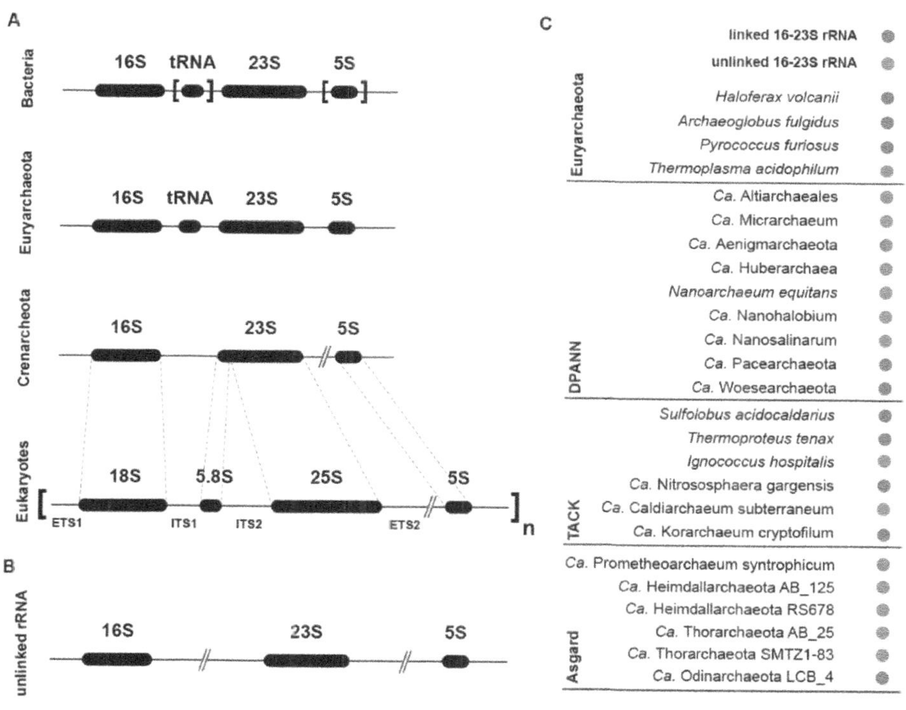

Figure 5.4. *Unlinked rRNA genes in archaea (reproduced from Jüttner and Ferreira-Cerca 2022). For a color version of this figure, see www.iste.co.uk/clouet/archaea2.zip*

COMMENT ON FIGURE 5.4.– *a) Typical rRNA genes organization in bacteria, archaea and eukaryotes. Typical rRNA genes transcriptional unit organization in bacteria, Euryarchaeota, Crenarchaeota and eukaryotes are depicted. The evolutionary relationship of the different rRNA genes is depicted by dashed lines.*

The brackets indicate that the presence and/or position of the 5S and/or tRNA within the rRNA genes varies. The 5S rRNA can, in some cases, be unlinked from the main transcription unit as indicated by //. ETS1: external transcribed spacer 1; ITS1: internal transcribed spacer 1; ETS2: external transcribed spacer 2; ITS2: internal transcribed spacer 2. b) Unlinked rRNA genes organization. Unlinked archaeal rRNA genes organization, originally described in T. acidophilum (Tu and Zillig 1982), is schematically depicted and the rRNA gene separation is indicated by the red //. c) Examples of linked and unlinked rRNA genes across archaea. Selected examples of linked and unlinked rRNA genes across archaea are depicted. Blue and brown dots indicate the presence of linked and unlinked 16S–23S rRNA genes, respectively. The presence of either linked or unlinked rRNA genes organization in different individual organisms classified as Aegnimarchaeota is indicated by a half blue/brown dot. Information on genome organization from the indicated organisms was obtained from NCBI, DOE-JGI or eLSMG and are provided in detail in Jüttner and Ferreira-Cerca (2022). Note the strong prevalence of unlinked 16S–23S rRNA in the newly discovered Asgard and DPANN superphyla, which further increase the so far recognized (Brewer et al. 2020) relative prevalence of unlinked rRNA genes in archaea. Note that the information provided does not reflect the overall quantitative prevalence of linked and unlinked rRNA genes across the archaea.

Most *Euryarchaeota* possess linked 16S-23S-5S rRNA genes separated by spacers sequences. In most cases, a tRNA (tRNA^{Ala}) gene also separates the 16S rRNA from the 23S rRNA (Figure 5.4). Moreover, in some *Euryarchaeota*, a second tRNA gene can be found downstream of the 5S rRNA gene. In contrast in *Crenarchaeota*, and likely in most members of the TACK superphylum, the 5S rRNA genes are physically separated from the two linked large rRNAs, whereas no tRNA is found between the latter two. Several exceptions to these archetype rDNA organizations can be found, for example, in the Euryarchaea *Thermoplasma acidophilum* or *Nanoarchaeum equitans*, but also commonly in the Asgard or *Diapherotrites*, *Parvarchaeota*, *Aenigmarchaeota*, *Nanoarchaeota* and *Nanohaloarchaeota* (DPANN) superphyla where the three rRNAs are physically separated from each other and transcribed separately.

Like in bacteria, the archaeal r-proteins are encoded within sometime very large (several kilobases) operons, scattered throughout the archaeal genomes.

Although similar to their bacterial counterparts, the polycistronic organization of the gene coding for the rRNA and r-proteins are presumably facilitating the stoichiometric production of the ribosomal constituents, regulation of the synthesis of the archaeal ribosomal constituents remains merely unknown.

5.1.4. *Ribosome biogenesis process*

In the following parts, we will focus on the molecular mechanisms liberating and further processing the precursor rRNAs into their mature functional form. Moreover, we will cover molecular principles of r-proteins assembly and how ribosome biogenesis factors putatively facilitate ribosomal subunit maturation. Finally, "decoration" of the rRNA by the covalent addition of RNA modifications will be the final aspect of ribosomal subunit biogenesis that will be covered.

5.1.4.1. *Ribosomal RNA processing*

Although some features of the rRNA pathway are reminiscent of the bacterial or the eukaryotic ones, the archaeal rRNA processing pathway(s) has its own molecular specificities, and it is tentatively summarized in Figure 5.5. It is assumed that the rDNA genes are transcribed as a long primary transcript comprising external and internal spacer sequences flanking the mature rRNA regions (see above). However, the existence of such long primary rRNA transcript, probably owing to its shorter half-life, has not been formally demonstrated in archaea. Similarly, in bacteria, and due to fast co-transcriptional processing events, the primary transcript is only observed in its full length in strain defective for the earliest rRNA maturation event.

In most prokaryotic organisms, the sequences flanking the two large rRNAs (16S and 23S rRNAs) have extended sequence complementarity enabling the formation of double helical stems, known as processing stems. Although in most bacteria these processing stems are substrates for the endonuclease RNase III, in archaea these stems typically contain a bulge–helix–bulge (BHB) motif that is recognized and cleaved by the tRNA splicing endonuclease (*endA*) (Figure 5.5). Akin to intron-containing tRNA maturation (see section 5.2.3), the resulting cleavage products are then covalently ligated, thereby generating archaea-specific circular-pre-rRNA intermediates (Figure 5.5). Recent studies have shown that the BHB motif and the tRNA splicing endonuclease *endA* are required for the formation of these circular-pre-rRNA intermediates and the downstream mature rRNA in vivo. Together these findings suggest a key role of pre-rRNA circularization for the maturation of functional ribosomal subunits. However, the exact functional contribution and advantage of these archaea-specific events for the ribosomal subunit biogenesis process remains to be fully understood. Interestingly, the presence of BHB motifs surrounding the 16S and 23S rRNA has been predicted in many archaea. However, these observations suggest that circular pre-rRNA formation may be widespread in the archaeal domain, experimental evidence will be necessary to verify the real prevalence of these pre-rRNA intermediates across the archaeal world.

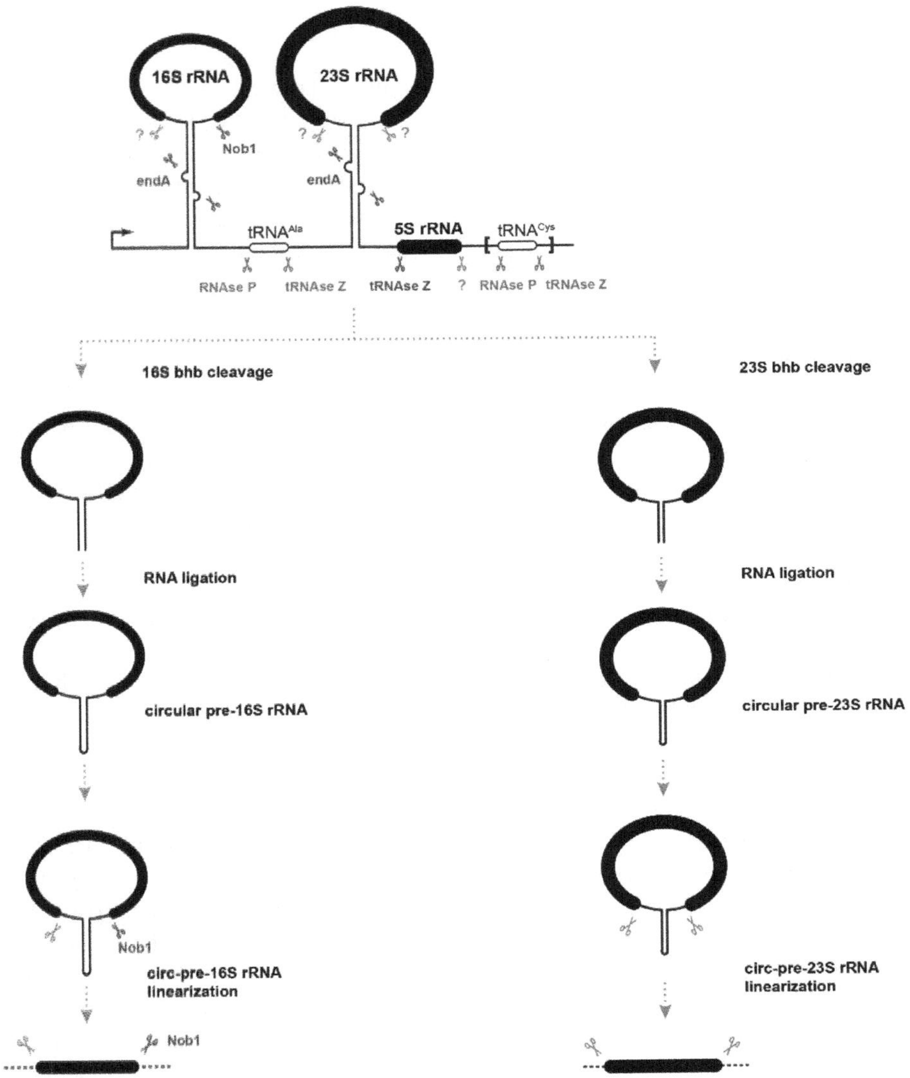

Figure 5.5. *Ribosomal RNA processing in the Euryarchaeon Haloferax volcanii. For a color version of this figure, see www.iste.co.uk/clouet/archaea2.zip*

COMMENT ON FIGURE 5.5.– *Ribosomal rRNA processing pathway in H. volcanii based on previous studies (Knüppel et al. 2018; Grünberger et al. 2020; Jüttner et al. 2020; Schwarz et al. 2020; Londei and Ferreira-Cerca 2021; Knüppel et al. 2021). Known ribonucleases involved in rRNA maturation are indicated: in red when in vivo/in vitro functional evidence is documented; in blue only shown in vitro; and in lilac, if predicted by functional analogy. Gray scissors indicate processing steps with unknown ribonucleases. In brief, the primary polycistronic transcript is processed by the tRNA splicing endonuclease endA at bulge–helix–bulge motifs present within the processing stems flanking the 16S and 23S rRNAs. After this step, both pre-rRNAs are circularized via ligation of the newly formed 5′and 3′extremities. The generated circular pre-rRNAs are a unique hallmark of archaeal ribosome biogenesis. During further maturation, the pre-rRNA is linearized and further processed to generate the mature 5′and 3′ends.*

After completion of yet unknown maturation steps, the circular pre-rRNAs are then further linearized and processed to their mature form presumably by the action of endo- and exo-nucleolytic activities, which in most cases still needs to be identified (Figure 5.5). So far, only the endonuclease Nob1 has been suggested to be able to mature archaeal 16S rRNA 3′end. However, these results are only based on in vitro studies, and by functional analogy to the Nob1 in vivo function described in eukaryotes.

5.1.4.2. *Ribosomal subunit assembly and factor facilitated ribosome assembly*

Ribosomal subunit assembly collectively refer to the steps by which r-proteins associate (assemble) to the pre-rRNA scaffold. Pioneering studies from the Nomura and Nierhaus laboratories have demonstrated that bacterial ribosomal subunits can be reconstituted in vitro solely using the isolated mature structural components, the rRNAs and r-proteins. These early experiments provided key insights into the auto-assembly nature of the ribosomal subunit production and that the 3D/4D structural information is inherently encoded within the primary sequences of these structural components. Remarkably, this feature seems to be conserved in archaea as well. In fact, pioneering work of the Londei laboratory demonstrated that highly evolutionary divergent archaeal ribosomal subunits could be reassembled from their isolated constituents in vitro. In stark contrast, in vitro assembly of eukaryotic ribosomal subunits solely from structural components was not described thus far. In fact, only one report of in vitro reconstitution of eukaryotic ribosomal subunits is documented. However, successful reconstitution was only possible in the presence of immature rRNA (pre-rRNA) and additional nucleolar fraction.

Finally, in vitro r-proteins omission experiments in bacteria demonstrated that r-proteins assembly proceeds via a stepwise hierarchical pathway whereby some r-proteins are binding independently of each other's to the rRNA, and others require prebinding of some specific r-proteins to the rRNA. In addition to such hierarchical assembly, r-proteins assembly is also cooperative, whereby interdependency (co-stabilization) is also seen. Remarkably, these general principles of r-protein assembly seem to be valid in vivo and may also apply to eukaryotic system and are likely to apply in archaea.

Although auto-assembly may be reminiscent of the ancestral nature of ribosomal subunit assembly, the conditions used in most of these in vitro reconstitution experiments are far to be mimicking the physiological conditions found in the respective organisms, thereby suggesting the presence of additional factors facilitating ribosome assembly in vivo.

While the search and discovery for these ribosome biogenesis factors has revealed a few dozens of factors in bacteria, genetic and biochemical purification of preribosomal particles in eukaryotic model organisms (e.g. yeast) has enabled to describe more than 200 ribosome biogenesis factors, most of which are eukaryotic specific. Among these factors, a significant fraction homes various enzymatic activity, mostly NTPase activity (ATPase, GTPase, RNA helicases, etc.), which may contribute to promote energy-dependent steps of the ribosomal subunit biogenesis process. Interestingly, although GTP-dependent processes are predominant in bacteria, ATP-dependent processes are strikingly more frequent in eukaryotes.

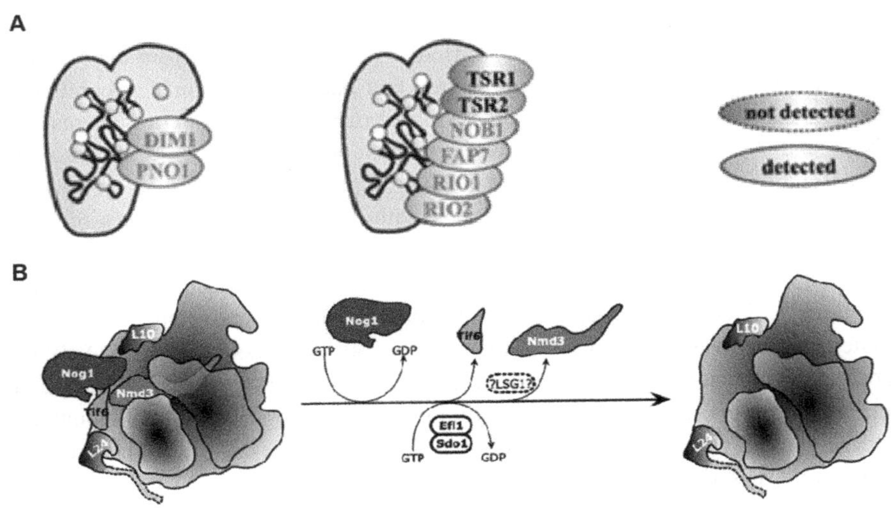

Figure 5.6. *Conservation between archaea and eukaryotes of the late steps of ribosomal subunits maturation (reproduced from Birikmen et al. 2021). For a color version of this figure, see www.iste.co.uk/clouet/archaea2.zip*

COMMENT ON FIGURE 5.6.– *a) Late steps of SSU maturation likely to be conserved between eukaryotes and archaea. The SSU yeast complexes involved in the final processing of the preribosomal subunits in the cytosol are shown together with the involved protein components. Factors without detectable counterparts in the archaea are shown as gray-shaded ellipses. Reproduced from Ebersberger et al. (2014). b) Late steps of LSU maturation likely to be conserved between eukaryotes and archaea. Shared factors are indicated. Note that there is no evidence for the presence of the ribosome biogenesis factor LSG1 in archaea (hatched oval).*

Importantly, it should be noted that the numbers of bacterial ribosome biogenesis factors are strongly reduced in comparison to eukaryotes. Moreover, most ribosome biogenesis factors are not conserved from bacteria to eukaryotes. In contrast, several sequence homologues of eukaryotic ribosome biogenesis are found in most archaeal genomes. However, in most cases their function for archaeal ribosome biogenesis remain yet to be explored in vivo. It is noteworthy that many of these shared eukaryotic-like ribosome biogenesis factors in archaea are known to be involved in the late steps of SSU and LSU biogenesis. To date, those which have been studied in archaea have similar function than their eukaryotic counterparts (Figure 5.6). Together, these observations suggest that at least some aspects of ribosomal subunit biogenesis in archaea may have similarity with the eukaryotic ribosome biogenesis pathway. Nonetheless, the search for archaea-specific ribosome biogenesis factors or the functional characterization of all eukaryotic-like ribosome biogenesis factors for archaeal ribosome biogenesis remains to be revealed.

5.1.4.3. *Ribosomal RNA modifications*

The presence of covalent RNA modifications is known since the end of the 1950s. The chemical modification process requires different molecular machineries and can occur co- and/or post-transcriptionally. Up to now, more than 100 different RNA modifications have been described. Interestingly, these RNA modifications expand and/or modify the chemical/structural properties of the canonical RNA nucleotide repertoire (AUGC). Due to their potential to regulate gene expression and the development of high-throughput technologies facilitating their discovery, RNA modifications analysis, also known as epitranscriptomics, has gained a tremendous new attention over the last decades. Although rRNA modifications have been explored in bacterial and eukaryotic systems in details, our understanding of the function and diversity of rRNA modifications in archaea remains, in comparison, in its infancy.

In this section, we will focus on the rRNA modifications found in archaea. In section 5.2, tRNA modifications, which are by far more diverse, will be the subject of our attention.

Herein the chemical nature, positions and diversity of rRNA modifications encountered in archaea will be described. Moreover, we will portray the key known type of molecular machineries involved in these modification reactions and that when their underlying molecular mechanisms are known. Finally, we will discuss the potential implications of these RNA modifications to control ribosome biogenesis and to establish functional ribosomal subunits.

5.1.4.3.1. Ribosomal RNA modifications in archaea: a short overview

rRNA modifications are present in all rRNA examined so far. However, their diversity (chemical nature, numbers and position) can be diverse across the tree of life. Two types of rRNA modifications are commonly found: (1) base modifications, that is, the nucleotide moiety is modified, and (2) ribose modifications, that is, the ribose counterpart is modified. Among these modifications, two sorts of rRNA modifications, ribose modification in the form of 2′O-methylation and isomerization of the uridine base into pseudouridine (pseudouridylation), are predominantly found across the tree of life.

In addition to these two major types of rRNA modifications, additional base modifications are also found. However, their presence and quantity vary between organisms. The commonly found additional modifications are base methylations (m^1, m^3, m^5, m^6A, $m^6{}_2A$.), but also acetylation (N^4ac) or larger type of modifications (e.g. acp^3/C*N330) (see Toolbox 5.2).

Overall, these rRNA modifications primarily cluster within the ribosomal subunit functional centers, for example, the A-, P- and E-site (see Chapter 7), and are believed to stabilize and/or fine tune the activity of the translation machinery. Interestingly, the exact position and/or chemical nature of these modifications is apparently flexible across the tree of life, suggesting that the functional contribution of the respective rRNA modification(s) in their respective structural environments prevails over their exact nature and/or relative position.

It is also important to note that the amounts of rRNA modifications strongly vary across archaea (10–100 modifications). Although halophiles show a very small amounts of rRNA modifications (*H. volcanii*, ~ 10), hyperthermophiles are characterized by a large increase of rRNA modifications and similar phenomenon is observed also for tRNA (see also section 5.2.4). Notably, it has been shown that *Thermococcales*, which are able to grow at high temperature and high-pressure environments, contain a large amount of additional rRNA acetylations, which in contrast to the majority of known rRNA modifications are scattered throughout the rRNA sequence. Overall, it has been suggested that the relative amount of rRNA modifications and their diversity may reflect organism-specific adaptation to their respective environmental condition and/or organism-specific evolutionary trajectories.

Finally, several studies have suggested that the rRNA modification stoichiometry may also change according to environmental variations, thereby enabling short-/long-term adaptations of the translation machinery. Although the functional consequences of this dynamic rRNA modification diversity for the translation process is not yet understood, by virtue of the almost infinite combinations of possible rRNA modifications, ribosomes appear to be potentially more heterogenous and dynamic than previously believed.

In the following, we examine the modification machineries involved in the RNA modification process and how do they recognize their respective rRNA substrates?

Two types of machinery can be distinguished: (1) stand-alone protein enzymes that are ubiquitously found across the different domains of life; and (2) RNA-guided modifications, which utilize ribonucleoprotein complexes to guide and modify the target rRNA in an RNA sequence-dependent manner. Importantly RNA-guided rRNA modifications are uniquely found in archaea and eukaryotes.

Due to their respective mechanism of action, the two types of RNA modification machineries generate, on the one hand, different molecular constraints to the ribosome biogenesis process and, on the other hand, ribosome biogenesis variability that may also influence the overall rRNA modification landscape.

In the following, we will illustrate molecular principles of stand-alone and RNA-guided modifications machineries in more details and highlight some principles of substrate recognition enabling the specific introduction of the corresponding rRNA modification.

5.1.4.3.2. Stand-alone enzymes

Stand-alone enzymes are the most basic unit carrying the RNA modification process and are found across the different domains of life. Conceptually, most of these enzymes recognize a cognate substrate with distinct enzyme-specific properties. As such, the molecular determinants enabling this specific recognition vary on a case-by-case basis. Accordingly, enzyme::substrate recognition can depend on a combination of RNA sequence/structure information or in other cases modification-competence depends on additional layer of structural information, which is established by the formation of discrete ribosomal subunit (local) assembly intermediates. In the latter, the topological information exerted by the rRNA sequence/folding and r-proteins content is integrated. Together, these different types of substrate recognition also suggest that rRNA modifications most likely occur within a restricted window of opportunity, and thus are timely regulated during the ribosome biogenesis process.

It is also interesting to note that only one stand-alone enzyme, the dimethyl-transferase KsgA/Dim1, is almost universally conserved throughout the tree of life, whereas all of the other stand-alone enzymes are restricted to either one or two domains of life (when considering the 3D topology of the tree of life).

5.1.4.3.3. RNA-guided modifications

In addition to stand-alone enzymes, archaea and eukaryotes utilize evolutionary related RNA-guided modification systems. RNA-guided modifications are of two types: 2′-O methylation of ribose and isomerization of the uridine base into pseudouridine. These two modifications are introduced by two distinct RNA-protein complexes, the C/D box- and H/ACA box-containing small RNA (sRNA).

Although the sRNA moiety specifies the sequence position of the target rRNA, the protein complexes contain either the methyltransferase (C/D-box) or the pseudouridine synthase (H/ACA box) activities. As such, the respective modifying enzyme activities are riding on a guide RNA, which by virtue of RNA duplex formation with the target rRNA exposes the substrate nucleotide toward the modifying enzyme catalytic center.

Archaeal research has been instrumental to our detailed molecular understanding of rRNA modification process by C/D and H/ACA box RNP. Early on, the superior biochemical properties of proteins derived from extremophile archaea were used to biochemically reconstitute active sRNP complexes and determine some of their structural and functional properties.

A few years after their discovery, the structure and general properties were analyzed with X-ray crystallography, cryo-electron microscopy and various biophysical and biochemical assays. From these studies, a general model of how these RNP complexes recognize and modify their respective substrate has emerged and is summarized in Figure 6.2 of Chapter 6.

Guide-RNA-dependent modifications rely primarily on rRNA::sRNA complementarity and may impose important physical and kinetic constraints to the ribosome biogenesis process and its evolution history. On the one hand, rRNA::sRNA duplex and the resulting topological constraint may delay rRNA folding and/or assembly of r-proteins. On the other hand, mechanisms unwinding the rRNA::sRNA must be established at the molecular level to facilitate release of the sRNP from its substrate nascent ribosomal subunit. In eukaryotes, most s(no)RNA-dependent modifications are presumably occurring very early during ribosome maturation, and unwinding of the rRNA::sRNA is supported by the action of ATP-dependent RNA helicases. In contrast, the timing of RNA-guided modifications and the RNA-helicases facilitating dissociation of the sRNP complex remain to be determined in archaea.

5.2. Synthesis and maturation of tRNAs

Transfer RNAs (tRNAs) are short noncoding RNAs that function in concert with ribosomes in all cells. Their main function is to provide the correct amino acid for the growing polypeptide chain. tRNAs are often called adaptor molecules because they assure the correct translation of the genetic code into protein sequence by matching the codon with the correct amino acid. Genes encoding tRNAs are transcribed in an immature form, called pre-tRNAs that need to be matured by concerted action of multitude of RNA-modifying and RNA-processing enzymes. In this section, we will describe how these maturation processes are conducted in archaea and highlight some specific examples for this domain of life. Despite being a distinct domain of life, the archaea are often characterized by a mixture of bacterial and eukaryotic features and, as we have already seen for ribosome biogenesis (see section 5.1), this is also the case for tRNA biogenesis pathways. Such a "hybrid" character of these essential pathways is a valuable asset helping us understand how the basic molecular machineries of present-day cells emerged in course of evolution.

5.2.1. *tRNA structure*

Mature tRNAs are among the smallest cellular RNA species with their size ranging between 70 and 100 nucleotides. They fold into a characteristic secondary structure, discovered in 1965, known as "cloverleaf" which is highly conserved in the three domains of life (Figure 5.7). For practical reasons, the numbering of ribonucleotides in tRNAs has been set from 1 to 76 and follows a universal rule. Four regions can be recognized in the secondary tRNA structure: the acceptor stem, a helix formed by the complementary 5′ and 3′ extremities and three stem loop structures namely the D-loop, anticodon loop and TΨC loop. In some tRNAs, a fourth stem loop region of variable size – the variable loop – is found between the anticodon loop and TΨC loop. The 3′ end of tRNAs is universally terminated with CCA triplet of nucleotides, to which one of the 22 proteinogenic amino acids is covalently linked. The anticodon loop carries the triplet of nucleotides at positions 34–36, which pair in antiparallel fashion with codons of translated mRNAs.

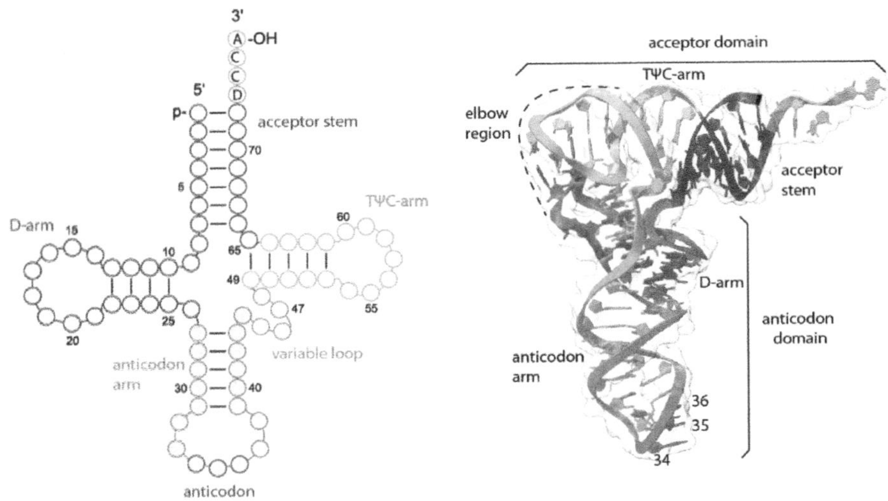

Figure 5.7. *tRNA structure. For a color version of this figure, see www.iste.co.uk/clouet/archaea2.zip*

COMMENT ON FIGURE 5.7.– *Cloverleaf structure of cytosolic tRNA^{Phe} from Saccharomyces cerevisiae is shown on the left with secondary structures indicated in different colors: acceptor stem (blue); TΨC-arm (yellow); variable loop (cyan); the anticodon arm (green); and the D-arm (red). The numbers correspond to the universal tRNA-numbering of nucleotides. The discriminator base (D) is depicted in orange and the CCA tail in pink. Several tertiary interactions (not shown) between spatially distant nucleotides are necessary to fold this molecule into L-shaped one shown on the right. The crystal structure of this tRNA^{Phe} from S. cerevisiae is deposited in the Protein Data Bank (entry code 1EHZ). The acceptor domain is composed of stacked TΨC-arm and acceptor stem, whereas the D- and anticodon arm form the anticodon domain. The region where the two domains interact with each other is called the elbow region. The white shadow corresponds to the surface representation of the same tRNA.*

The different stem-loop structures interact with each other, through several key intramolecular pairings, to form a three-dimensional L-shaped structure. One end of this structure corresponds to the anticodon arm, and the other end corresponds to the acceptor arm, while the D, TΨC and variable stem-loop regions form the "elbow" between the two arms (Figure 5.7). This conserved general structure determines the recognition of tRNAs by the ribosome, elongation factors as well as by various enzymes (notably the tRNA maturation enzymes, which we will describe below), thus its preservation is essential for tRNA function.

5.2.2. *Diversity of tRNA genes*

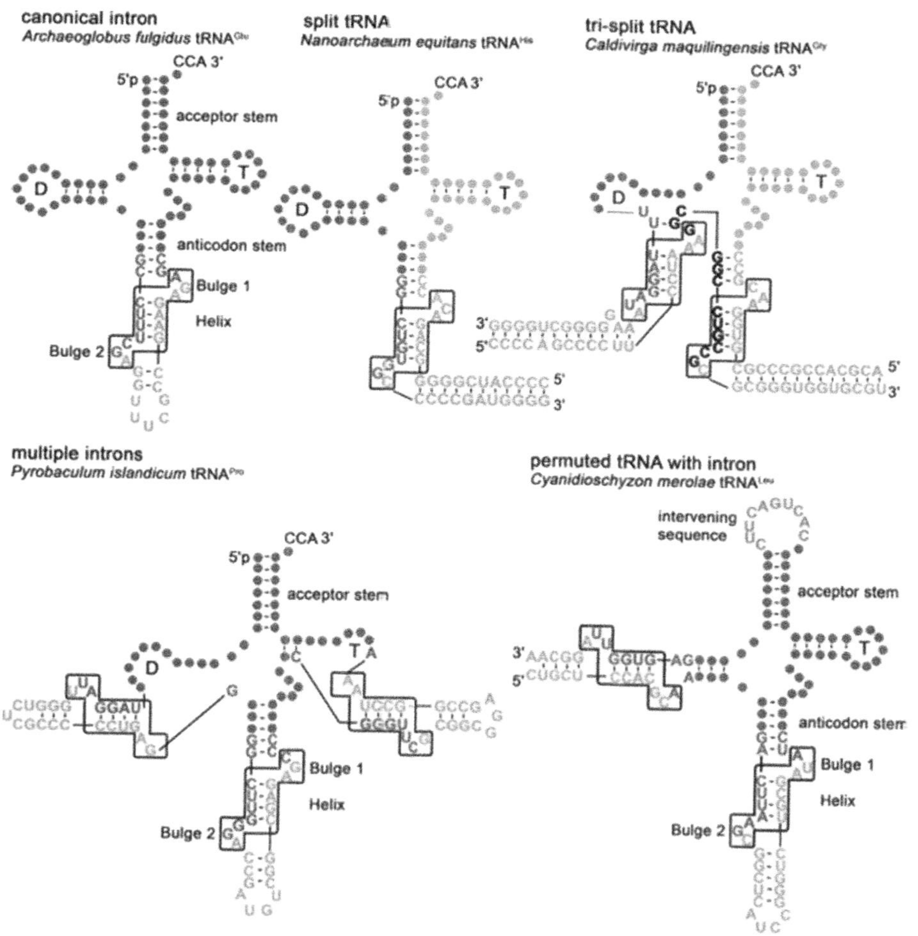

Figure 5.8. *Examples of unusual archaeal tRNA genes (figure reproduced from Heinemann et al. 2010). For a color version of this figure, see www.iste.co.uk/clouet/archaea2.zip*

COMMENT ON FIGURE 5.8.– *A classical version of pre-tRNA with a single intron sequence (in orange) is shown on the top left. Different shades of blue indicate the assembly of the pre-tRNA from different transcripts. For example, the tRNA^{His} of Nanoarchaeum equitans is stitched together from two distinct transcripts, the genes encoding these tRNA halves can even be located at distant loci in the genome. The two tRNA halves have complementary sequences that allow us to match the two halved together. The tRNA^{Gly} from Caldivirga maquilingensis is an even more extreme case since its tRNA gene is split into three separate parts. Multiple introns have also been detected in archaea as shown for the tRNA^{Pro} of Pyrobaculum islandicum. The permuted tRNA genes, such as one encoding tRNA^{Leu} in the red algae Cyanidioschyzon merolae, code the 3′ half of the tRNA upstream from the 5′ half, and the two parts are connected with an intervening sequence (green). Helix–bulge–helix motifs (boxed) are recognized by the tRNA splicing endonuclease called EndA, which cleaves the intron sequences from the tRNA exons (in blue).*

Every archaeal cell carries a set of genes encoding all necessary tRNAs, which are expressed either as individual genes or as polycistronic operons together with other tRNAs, rRNAs or mRNAs. The number of tRNA genes per genome is variable among archaea but, typically, each species encodes between 30 and 50 tRNA genes.

In addition to canonical single transcript tRNA genes, archaea distinguish themselves by several unusual tDNA arrangements, some of which are shared with eukaryotes. Presence of introns in tRNA genes is quite common in archaea and eukaryotes and will be presented in more detail in section 5.2.3.3. More exotic arrangements are found almost exclusively in thermophilic archaea and, rarely, in red and green algae. These unusual genes are split into two or even three segments or are circularly permuted, whereby the 3′ half of the tRNA gene precedes the 5′ half in an operon. In either case, the generation of full-length tRNA depends on the joining of tRNA fragments by annealing of complementary noncoding sequences, which form particular secondary structure called BHB (Figure 5.8). Similar to rRNA processing (see section 5.1.4.1), this structure is recognized and processed by e*ndA* endonuclease to yield tRNA parts devoid of adjacent sequences.

Finally, a specific ligase named RtcB intervenes to join the tRNA fragments into complete cloverleaf molecule (see also section 5.2.3.3). Split tRNA genes of hyperthermophilic crenarchaeon *Caldivirga maquilingensis* are an interesting case of this phenomenon because the tRNA fragments carrying the anticodon can be combined with other fragments similar to jigsaw puzzle to create synonymous tRNAs.

5.2.2.1. *Maturation of tRNA*

tRNA surgery

The fabrication of mature functional tRNAs resembles a complex multi-step surgery where a patient, a precursor pre-tRNA, is cut and then stitched together by several specialized nano-sized surgeons. The operation begins by slicing out, at a precise spot, the extra "limbs", which are the two extremities in between which the tRNA sequence is positioned. Quite often, the tRNA molecules have another "internal extra limb", which splits the tRNA sequence into two parts. The removal of this "limb" requires another surgery called splicing. However, this operation puts the patient tRNA molecule in a bad condition since the molecule is split into two halves afterwards. Luckily, the cellular hospital has another specialist surgeon that can heal the tRNA by stitching together (ligating) the two halves. Now the patient looks like themself, meaning that it can resume its typical cloverleaf shape. However, this heavy surgery is not sufficient for tRNA to be operational for work in protein synthesis. Actually, several plastic surgeons need to intervene and "embellish" the tRNA molecule. One of these surgeons will add an extra triplet of nucleotides with the precise sequence CCA on one extremity of the tRNA. Other plastic surgeons will modify some of the nucleotide bases and this will help the tRNA to become more stable and more accurate in decoding the genetic code. The last, but not least, operation step is more of a coronation and less of a surgery. The ceremony involves the attachment of one of the 22 proteinogenic amino acids to the tRNA "head". By "embracing" its tRNA in a molecular dance, the crown deliverer (tRNA synthetase) can recognize its own tRNA queen and attach the correct amino acid. Only then the tRNA queen can take part in the molecular ballet staged by the ribosome. In the following text, we will describe in more detail the cutting, splicing, healing, decorating and crowning operations and the different nano-surgeons involved.

5.2.3. *Transcript processing*

In this section, we will describe how the precursor tRNA transcript is processed to produce a mature tRNA. A typical pre-tRNA transcript contains 5′ leader and 3′ trailer sequence, and in many cases, one or even more introns split the tRNA sequence into two or more parts. These additional sequences need to be removed by endonucleases from the pre-tRNA transcripts and the resulting exons are ligated together to produce the canonical cloverleaf tRNA molecules. In most cases, the CCA triplet of ribonucleosides needs to be added enzymatically to the 3′ extremity of tRNAs. Full sequence tRNAs are then "decorated" by a variety of nucleoside modifications that require the action of numerous tRNA modification enzymes. Different steps of tRNA maturation process and the involved enzymes are summarized in Figure 5.9 and will be described briefly in the following sections. Bear in mind, however, that the nature and the order of tRNA processing events as shown in Figure 5.9 was deduced from the work in model archaeon *Haloferax volcanii* but, as we have seen in section 5.2.2, archaea code a multitude of pre-tRNA variants, each requiring specific maturation pathways.

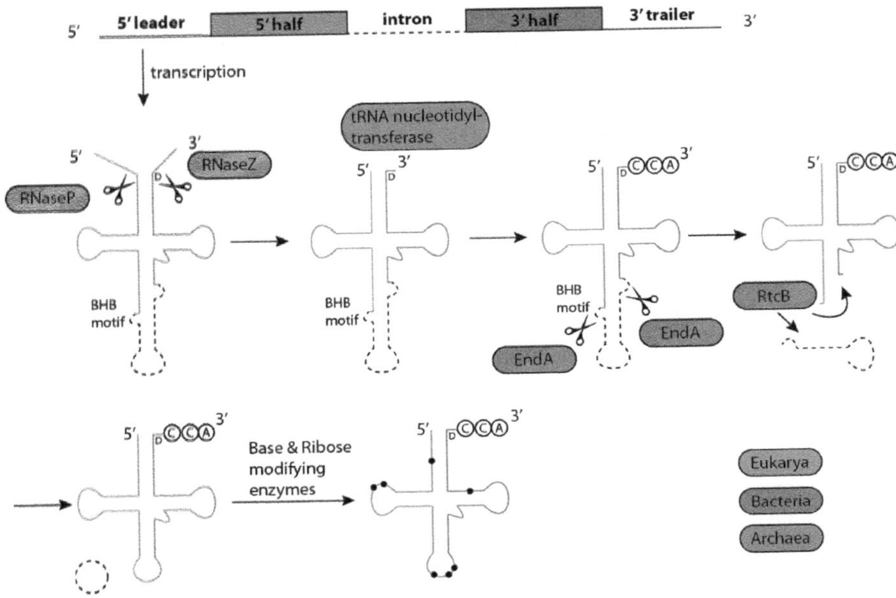

Figure 5.9. *Overview of the tRNA maturation process in archaea and associated enzymes (figure adapted from Clouet-d'Orval et al. 2018). For a color version of this figure, see www.iste.co.uk/clouet/archaea2.zip*

COMMENT ON FIGURE 5.9.– *On the top of the figure is shown the structure of a typical archaeal tRNA encoding gene composed of two exons (blue and red), an intron and 5' and 3' extensions. After transcription, the tRNA adopts its classical L-shaped structure. Here, the molecule is depicted as cloverleaf for simplicity. Two BHB structures are formed at the exon–intron junctions. As the first step of processing, two universal exonucleases RNaseP and RNaseZ cleave the 5' leader and 3' trailer sequences, respectively. Then, an archaea-specific tRNA nucleotidyl-transferase adds the CCA tail to the 3' end of the molecule. Next, the EndA endonuclease conserved in archaea and eukaryotes recognizes the BHB motifs and removes the intron sequence. Following the cleavage, the two halves need to be stitched together and this is done by the RtcB ligase, which is shared between archaea and eukaryotes. This enzyme also catalyzes the intramolecular ligation reaction resulting in circularization of the cleaved-off intron. Finally, some nucleotides will be chemically modified (black dots) by various tRNA-modifying enzymes. The sequence of processing events depicted here corresponds to that of the model archaeon Haloferax volcanii.*

5.2.3.1. *Processing of 5′ and 3′ extremities*

A unique archaeal RNA polymerase (RNAP) related to the eukaryal RNA polymerase II (RNAPII) is in charge of transcribing pre-tRNAs from corresponding genes. The pre-tRNA transcript processing begins with the action of two universal enzymes, RNase P and RNase Z, which are housekeeping endonucleases that carry out the internal cleavage of phosphodiester bond to remove 5′ and 3′ leader and trailer sequences, respectively.

RNase P is a site-specific endonuclease that cleaves the precursor tRNA at the junction between the single stranded 5′ lead sequence and the double stranded RNA formed by the acceptor stem of the tRNA. The presence of particular structural characteristics, such as canonical 3D structure of tRNA, and particularly its "elbow" are critical for the recognition of the pre-tRNA as a substrate and correct positioning of the cleavage site.

Typical archaeal RNase P holoenzyme is a ribozyme forming a ribonucleoprotein complex with at least five protein subunits assembled together with the RNA molecule that contains the catalytic activity. The RNA subunit of archaeal RNase P is globally similar in sequence and structure to bacterial RNase P but the associated protein subunits are homologous to counterparts in eukaryotic RNase P, thus illustrating the mosaic evolutionary origin of the archaeal RNase P complex (Figure 5.10).

Three types of RNase P exist in archaea. The A-type is the most widely spread, the M-type is present only in *Methanococci* and *Thermoproteales* and the recently discovered T-type is restricted to *Thermoproteaceae* and related crenarchaeal species. The basis for this differentiation is the structure of the catalytic RNA subunit. In A-type RNaseP, the RNA molecule alone is sufficient for substrate recognition and catalysis, but the presence of protein subunits is required for optimal activity. Interestingly, in addition to ancestral RNA-based form of the RNase P, a nonhomologous protein-only RNase P called HARP was recently detected in many *Euryarchaeota* and TACK group archaea and its 5′ pre-tRNA processing activity was reported for *H. volcanii* and *M. mazei Euryarchaeota*. However, the genetic evidence revealed that these HARP orthologs are not major contributors to 5′-end tRNA maturation and may have a (still) unknown specialized function in (t)RNA metabolism.

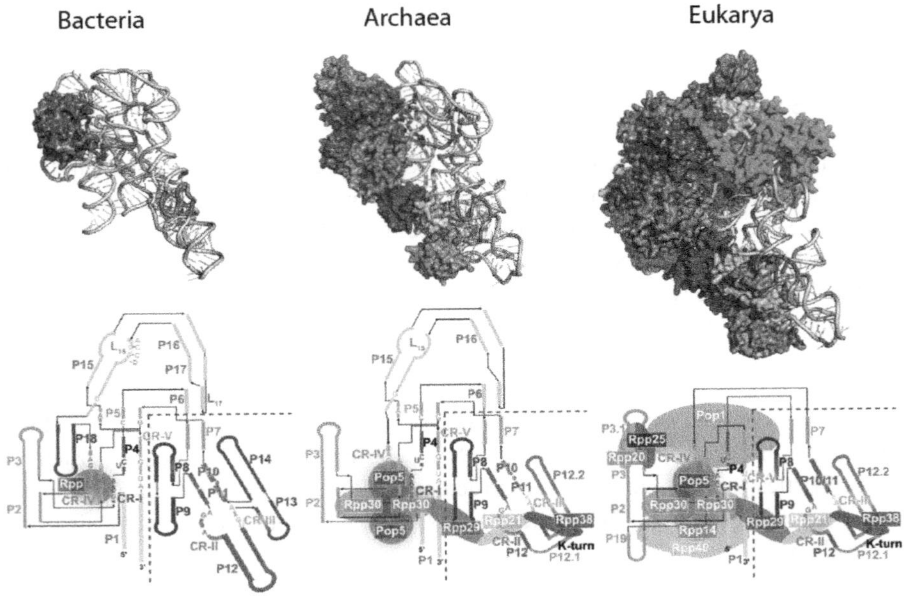

Figure 5.10. *RNaseP in the three domains of life (figure reprinted from Wu et al. (2018) with permission from Elsevier). For a color version of this figure, see www.iste.co.uk/clouet/archaea2.zip*

COMMENT ON FIGURE 5.10.– *Structure of representative RNaseP enzymes is shown at the top. Archaeal RNaseP is composed of an RNA molecule (shown in gray) that carries the catalytic center and five associated proteins (Pop5, Rpp30, Rpp21, Rpp29 and Rpp38). The homologous protein components have the same color. The bottom panel shows the secondary structure diagram of the RNaseP RNA component and associated proteins. Comparison of the RNaseP structures shows that the archaeal enzyme is an evolutionary intermediate between the RNA-dominated bacterial form and protein-rich eukaryotic version. The RNaseP thus evolved by gradually replacing RNA-based parts by proteins while maintaining a conserved catalytic RNA core structures.*

RNAse Z enzymes are universal metallo-dependent beta-lactamases, which cleave the 3′ trailer sequence of precursor tRNA. Archaeal RNase Z (aRNase) are polypeptides of about 250–350 amino acids that are present in almost all archaea. They require Zn^{2+} ions as co-factor and function as homodimers. Crystal structure of an RNase Z has not been reported but yet sequence analyses and computer models show presence of beta-lactamase domain and a flexible arm named exosite, which is used to bind mainly to T-arm and acceptor stem of pre-tRNA substrate. The cleavage produces a tRNA molecule with a free 3′ hydroxyl group. Most archaea do not code CCA triplet in the 3′ trailer of their tRNA genes and this essential part of each tRNA needs to be added in a separate reaction catalyzed by a dedicated nucleotidyl-transferase (see section 5.2.3.2. below). In this, archaea are similar to eukaryotes, but are different to bacteria, which typically encode CCA triplet in their pre-tRNAs. In archaea, the presence of this triplet in tRNA sequence was shown to inhibit the aRNase Z cleavage activity. This makes perfect sense because it prevents occurrence of endless cycles of CCA addition and removal with potentially disastrous outcome for the cell viability. But the situation is not always straightforward. The crenarchaeon *P. aerophilum* and the euryarchaeon *M. jannaschii* harbor a mixed population of pre-tRNAs, some of which encode the CCA triplet but are nevertheless recognized as substrate by aRNase Z. Even more intriguing is the case of *P. furiosus* and *A. pernix*, which encode only CCA-containing tRNAs. How the competing aRNase Z and nucleotidyl-transferase activities are regulated in these organisms to achieve proper 3′ pre-tRNA processing is currently unknown.

5.2.3.2. *Addition of CCA tail*

Now that we have described how the 5′ and the 3′ ends of pre-tRNA transcripts are usually (and in some cases unusually) processed we can go to the next step, which is handled by a special nucleotidyl-transferase which is also called CCA-adding-enzyme or tRNA terminal transferase. Archaeal tRNA terminal transferases are structurally distinct from bacterial and eukaryotic ones, even though they share a homologous N-terminal catalytic core. The 3′ nucleotidyl activity of these enzymes is occurring without template so how do these enzymes “know” that after adding two Cs an A needs to be added? An answer to this question came from the study of the *A. fulgidus* tRNA terminal transferase. Indeed, the specificity of the nucleotide-binding pocket evolves with the growing 3′ terminus such that after addition of two Cs the specificity switches from CTP to ATP for the final nucleotide incorporation (Figure 5.11).

In addition to its essential role of generating CCA tails to enable, as we will see later, the aminoacylation of tRNAs, the tRNA terminal transferases from the three domains of life were shown to catalyze the addition of the “double CCA tail” (CCACCA sequence) in unstable tRNAs. This prevents the aminoacylation of such destabilized tRNAs and marks them for degradation. The magnitude by which this tRNA surveillance mechanism contributes to the general translation quality control in vivo is difficult to establish because of the multiplicity of factors susceptible to destabilize the tRNA structure including tRNA modifications, which will be presented in section 5.2.4.

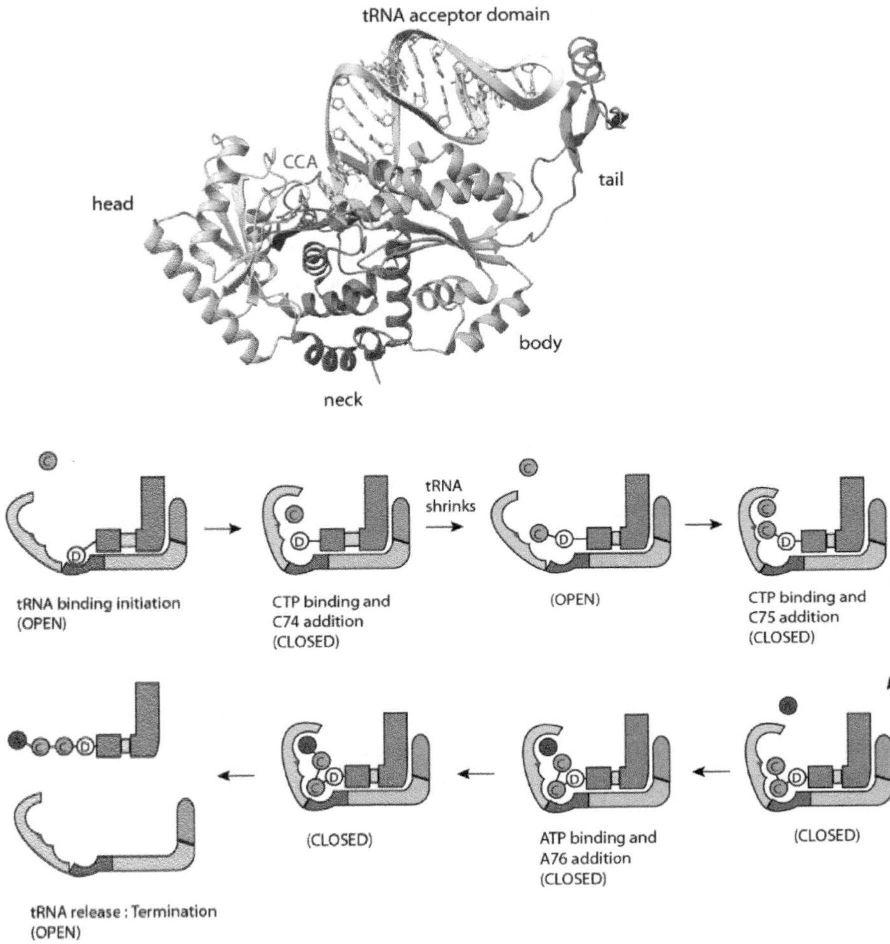

Figure 5.11. *Mechanism of CCA tail addition by an archaeal tRNA terminal transferase (figure adapted from Tomita and Yamashita 2014). For a color version of this figure, see www.iste.co.uk/clouet/archaea2.zip*

COMMENT ON FIGURE 5.11.– *At the top of the figure, we can see the crystal structure of CCA-adding enzyme from the archaeon Archaeoglobus fulgidus (PDB entry 2DRB). The protein contains four domains (head, neck, body and tail), whereby the head domain contains the catalytic site. The protein is bound to an RNA minihelix (in gray) corresponding to acceptor- and T-arm of tRNA. The CCA tail is indicated in orange color. The lower part of the figure explains how this enzyme works. The mechanism was revealed by comparison of several crystallographic structures of this enzyme in complex with various RNA primers with or without nucleotides. The enzyme and the tRNA molecule are represented schematically. The domains are colored using the same color code as above. The active site is indicated by a red triangle and D stands for discriminator nucleoside. The model highlights changes in the orientation of the head domain and the 3′ end of the tRNA during CCA addition. The open form of the enzyme is inactive. The binding of a CTP molecules into the active site leads to transition of the enzyme to a closed conformation, which catalyzes the addition of the C74 to tRNA. This triggers opening of the enzyme and shortening of the tRNA molecule such that the active site can accommodate another CTP molecule. tRNA is then further elongated by the addition of C75. Next, the 3′ extremity of tRNA refolds to make place for the entry of one ATP molecule and A76 addition proceeds without the open to closed conformational transition of the enzyme. Finally, the enzyme adopts open conformation and CCA-tailed tRNA is released.*

5.2.3.3. *Intron splicing and exon ligation*

In addition to some unusual gene arrangements described in section 5.2.2, archaeal tRNA genes often contain introns. In average, about 15% of archaeal tRNAs contain introns with variable sizes ranging from 11 to 129 nucleotides. In some archaea such as *Crenarchaeota* and *Thermoproteales*, the prevalence of introns in tRNA genes is high and reaches 48% to almost 70%, respectively. The presence of multiple introns is not rare and was recorded for 17% of intron-containing tRNA genes. tRNA introns are not archaea-specific feature but also occur in bacteria and eukaryotes. In these two domains of life, the tRNA introns are inserted between nucleotide 37 and 38 of the anticodon loop but in archaea more than 30 alternative positions were described throughout the entire tRNA molecule (Figure 5.12). What allows the archaea such promiscuity with regard to intron location? The answer lies in the particular features of the archaeal splicing mechanism. First of all, unlike bacteria, archaea do not use self-splicing introns but require a special tRNA splicing endonuclease called *endA,* which has partial homology to eukaryotic counterpart. Second, the archaeal *endA* recognizes the specific structural motif called BHB as necessary and sufficient determinant for splicing site, which explains why a particular location of the intron in the tRNA is not required (Figure 5.12).

As the result of *endA* action, two tRNA halves (3′ half and 5′ half) are produced and the intron is released. Now, the two halves need to be joined to obtain a canonical cloverleaf tRNA molecule. In archaea and eukaryotes, this is achieved by the action of RtcB ligase, which catalyzes the formation of the internucleotide bond that joins the two halves (Figure 5.9). RtcB shares no sequence or structural similarity with canonical nucleic acid ligases and, in marked contrast to those classical ligases, it relies upon GTP and Mn^{2+} for catalysis. In many archaea, the gene encoding RtcB is found in an operon with the gene encoding a small (~ 16 kDa) protein called archease. This protein was found to improve the efficiency of the RNA ligation and alter the NTP specificity of an archaeal RtcB in vitro such that the ligation proceeds efficiently with both ATP and GTP. Archease is widely conserved (but not ubiquitous) and was found to be critical for tRNA splicing in animals. However, the function of the archaeal archease in vivo remains to be investigated.

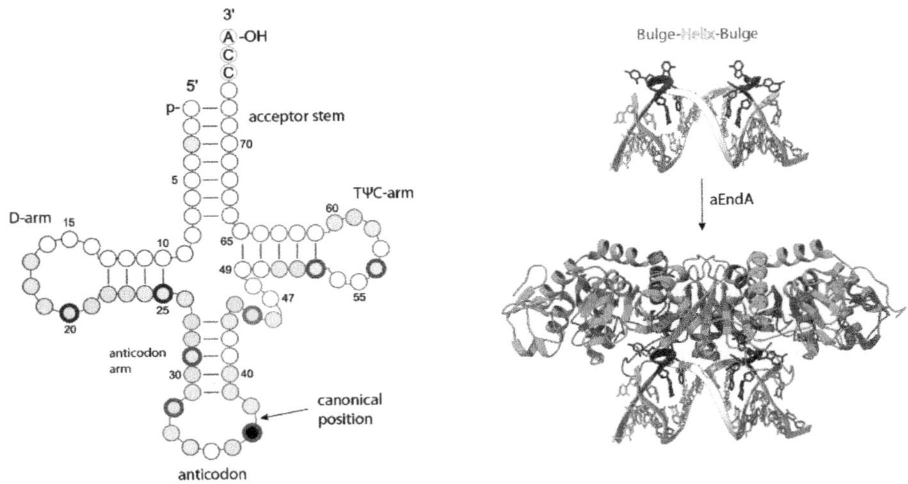

Figure 5.12. *Intron splicing in archaea (left panel adapted from Clouet-d'Orval et al. 2018). For a color version of this figure, see www.iste.co.uk/clouet/archaea2.zip*

COMMENT ON FIGURE 5.12.– *The left part shows the reported insertion positions of archaeal introns mapped on a cloverleaf structure of tRNA. Numbers correspond to universal numbering of tRNAs. Filled circles indicate the nucleotide located 5′ to the insertion site (including the sites found in split tRNA genes). Positions conserved across the phyla are colored in light orange, and those found only in Thermoproteales are in light blue. Thick circles indicate the positions with high frequency of insertion (more than 10). The right part shows the crystal structure of EndA from the archaeon Archaeoglobus fulgidus in complex with BHB motif (PDB entry 2GJW). Two EndA proteins are arranged in an antiparallel fashion and form a dimer. The structure further shows that the three nucleotides in the bulge are flipped out of their usual stacking position and interact with the catalytic site of each EndA protomer.*

5.2.4. *Nucleotide modifications*

We have seen so far how the precursor tRNA transcript containing the canonical four nucleotides (C, A, U and G) is processed to obtain a typical cloverleaf molecule ending with CCA tail. In this section, we will learn that, in addition to the four canonical nucleotides, all tRNAs carry chemically modified nucleotides that are synthetized posttranscriptionally by a variety of standalone or multisubunit enzymes. The world of tRNA modifications is so vast that it would probably require a book of its own to describe it entirely. Therefore, in this section we will briefly present some general information about the diversity, distribution and function of tRNA modifications. Then, we will describe, as an example, the synthesis and function of two tRNA modifications specific for archaea.

Let us start by saying that the tRNA molecules are the champions when it comes to modified nucleotides. There are almost ten times more modified nucleotides in tRNA (average ~ 17% of the total residues are modified) than in the other abundant RNAs such as rRNA (average 1–2% of total residues are modified). Moreover, at least 93 modifications are found in tRNAs out of over 130 structurally characterized modifications found across all types of RNAs. Their chemical complexity varies from simple methylation reactions to so-called "hypermodifications" whose synthesis often requires a whole cascade of enzymatic reactions. Although the exact positions that are modified vary between tRNAs from different species, the common pattern of modification, with a high density of modifications observed in the anticodon stem–loop and much less in the aminoacyl-acceptor stem, is conserved in most organisms (Figure 5.13).

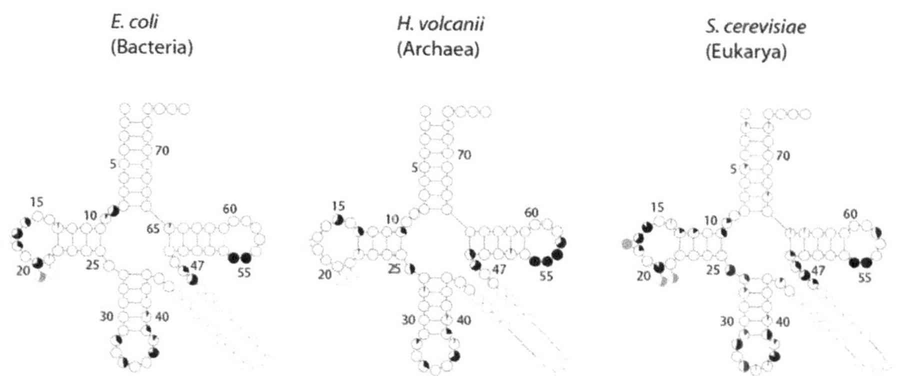

Figure 5.13. *Occurrence of modified nucleosides in tRNA (images courtesy of the MODOMICS database)*

COMMENT ON FIGURE 5.13.– *The figure shows cloverleaf depiction of tRNA molecules from three model organisms. The numbers correspond to universal numbering of tRNA nucleosides. The positions outside this numbering are depicted using gray circles. For each position, the frequency of modified nucleotides is indicated as pie chart. For example, position 55 (full black circle) is modified in all tRNAs from the three organisms.*

In general, more than half of the modifications found in tRNAs are domain specific, and only one fifth is universally distributed and another one fifth is found in only two of the three domains of life (Figure 5.14). tRNAs of archaea investigated so far contain at least 47 different tRNA modifications, and 16 of them, including archaeosine and agmatidine (see below for more detail), are specific to the archaeal domain.

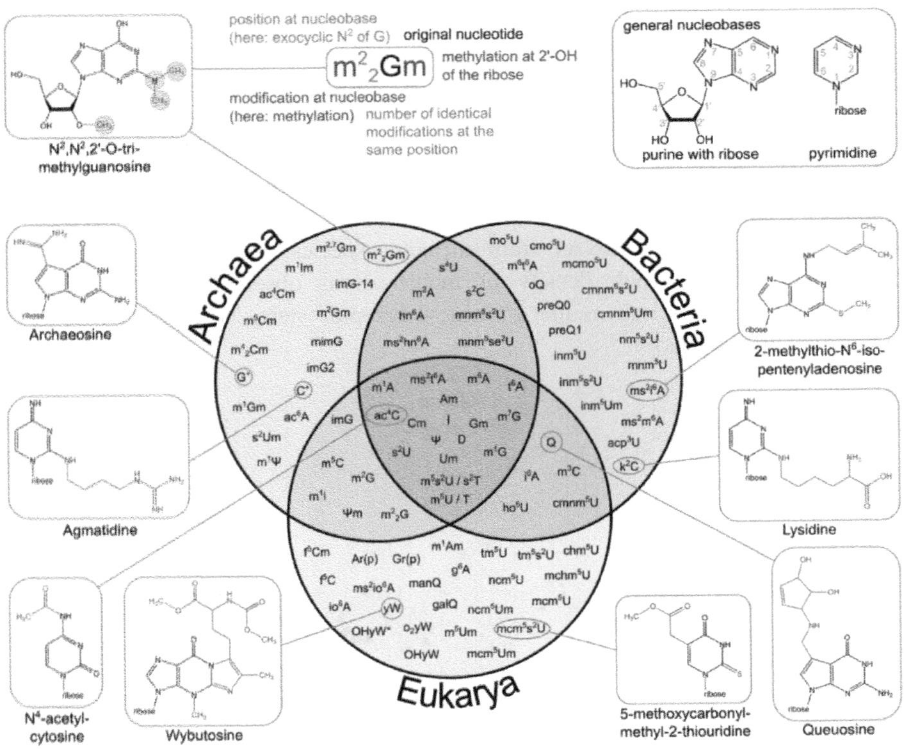

Figure 5.14. *Modified nucleosides found in tRNA and their distribution among living organisms (figure reproduced from Lorenz et al. 2017). For a color version of this figure, see www.iste.co.uk/clouet/archaea2.zip*

COMMENT ON FIGURE 5.14.– *Each modified nucleoside is denoted by a particular abbreviation that indicates the type of chemical modification and its position. The explanation on how to read these abbreviations is given in the upper left corner. General structure of nucleobases and the atom numbering is given in the top right corner. The Venn diagram in the middle shows the distribution of 93 different modified nucleosides in the three domains of life. Most of the chemical modifications, such as methylations, are quite simple but a few are much more complex. The chemical formula of some of them is shown whereby the canonical base is drawn in black and the modification in red.*

It is important to stress here that this is a very incomplete picture since the immense majority of tRNA sequences remain uncharacterized. Indeed, the identification and mapping of tRNA modifications is technically challenging and requires direct inspection of tRNA molecules using biophysical, biochemical and genetic approaches. For archaea, the systematic survey of the tRNA modifications repertoire exists only for one archaeal model species, *Haloferax volcanii* (Figure 5.15).

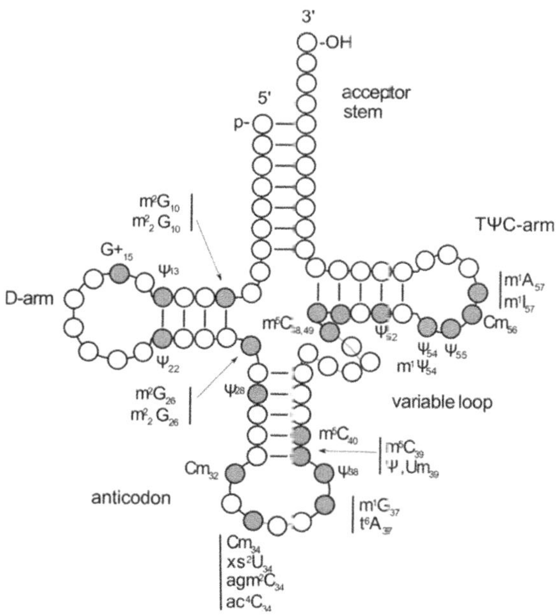

tRNA Modification		Experimentally validated in		
POSITION	GENE	*H. volcanii*	another archaea	another domain
m^2G10 / m^2_2G10	Trm-G10 HVO_0156		•	
Ψ13	Pus7 HVO_0658		•	
G^+15	TgtA1 HVO_2001	•	•	
	FolE2 HVO_2348	•		
	QueC HVO_1716	•		
	QueE HVO_1717	•		
	QueD HVO_1718	•		
	ArcS HVO_2008	•	•	
Ψ22	Unknown gene			
m^2G26 / m^2_2G26	Trm-1 HVO_0236		•	
Ψ28	Unknown gene			
Cm32	Trm-H HVO_2906			•
Cm34	Guide C/D sRNA	•		
xs^2U34	Trm9 HVO_1032			•
	ELP3 HVO_2888			•
	Tuc1 HVO_0580			•
agm^2C34	TiaS HVO_0339		•	
ac^4C34	TmcA HVO_2736	•		
m^1G37	TrmS HVO_0929		•	
t^6A37	YrdC HVO_0253		•	
	Kae1 HVO_1895			•
Ψ38/39	Pus3 HVO_1852	•		•
Um39	Guide C/D sRNA	•		
$m^5C40/39$	Trm4 HVO_1594	•	•	
$m^5C48/49$	Trm4 HVO_1594	•	•	
Ψ52	Unknown gene			
Ψ54/55 / m1Ψ54	Pus10 HVO_1979 / COG2376HVO_1989	•	•	
Cm56	Trm56 HVO_1173		•	
m^1A57/m^1I57	Trm1 HVO_1383		•	

Figure 5.15. *tRNA modifications identified in the archaeon H. volcanii and predicted and validated tRNA modification genes (figure adapted from Phillips and de Crécy-Lagard 2011). For a color version of this figure, see www.iste.co.uk/clouet/archaea2.zip*

COMMENT ON FIGURE 5.15.– *The identified modified nucleosides are mapped onto the cloverleaf tRNA molecule and the universal tRNA numbering is used for each position. Filled circles indicate that a modification was found at that position. Different modifications are denoted with a corresponding abbreviation (see Figure 5.14). The genes encoding predicted or validated modification enzymes are indicated in the table. Those that have been deleted in H. volcanii are in red.*

Despite this important progress, the archaeal organisms still lag behind *Escherichia coli* and *Saccharomyces cerevisiae* in terms of genetic characterization and in vivo function of tRNA modifications. More generally, there are only few representative organisms for which a significant number of tRNA molecules was characterized. On the tRNAmodviz database, there are currently 61 entries for tRNA sequences from archaea, 134 from bacteria and 390 from eukaryotes. Nevertheless, the recent progress in sensitive and high-throughput mass spectrometry and RNA sequencing technologies should push forward the mapping and identification of tRNA modifications in the years to come (see Toolbox 5.2).

As a general principle, the tRNA modifications fall into two categories depending on how they affect tRNA function: those that affect overall structure of the tRNA and those that target the functional centers of the tRNA, namely, the anticodon loop and/or sequences important for aminoacylation. The second category therefore has direct effect on decoding and protein synthesis. To illustrate how tRNA modifications can affect tRNA structure and translation, we present in the following text two examples of tRNA modifications that are found exclusively in archaea.

Agmatidine (C^+) is one beautiful example of how tRNA modification can modify the translation outcome. This hypermodification corresponds to the covalent attachment of agmatine to the wobble nucleotide C34 in $tRNA^{Ile}{}_{AUA}$ by a recently discovered TiaS enzyme (see Figure 5.14 for chemical structure). The purpose of this modification is to decode AUA initiation codon (which normally specifies an isoleucine) into methionine. To achieve this, following the formation of agmatidine, the $tRNA^{Ile}{}_{AUA}$ is not recognized any more by the isoleucyl-tRNA synthetase but instead becomes the substrate for the methionyl-tRNA synthetase. This modification is also an interesting example of how evolution can lead to two original solutions for the same problem. In bacteria, the reassignment of AUA initiator codon into methionine is achieved by addition of lysidine to the C34 $tRNA^{Ile}{}_{AUA}$ by an enzyme TilS, which has no sequence or structural homology with TiaS.

Archaeosine (G^+) is among the most complex modifications known to occur in tRNA (see Figure 5.14 for chemical structure). It is found in almost all archaea at position 15 of tRNAs, a site not modified in any tRNA outside the archaea. This is a chemically complex and singular modification characterized by an unusual 7-deazaguanosine core structure with a formamidine group at position 7. Accordingly, the synthesis pathway of Archaeosine is very complex and requires sequential action of seven different enzymes. Interestingly, the first part of the pathway leading to the synthesis of $preQ_0$ precursor (7-cyano-7-deazaguanine) is homologous in bacteria and archaea but the enzyme that catalyzes the attachment of $preQ_0$ to tRNAs is domain specific. The archaeal version of this enzyme is called TransGlycosylTransferase (arcTGT) and is encoded by the *tgtA* gene, which is highly conserved in archaea and is one of archaeal specific marker genes. G^+ is located at the interface of the D-loop/T-stem of tRNA and its predicted function is in stabilizing the overall tRNA structure via attractive forces between the positively charged formamidine group and the backbone phosphates in the vicinity. In line with this function, *tgtA* was identified as important for thermophilic lifestyle but this is not the only tRNA modification that thermophiles have in their arsenal, as we will see in the next paragraph.

Indeed, thermophiles and more generally extremophiles are exposed to environmental conditions, which are highly denaturing for their macromolecules. RNA is especially sensitive to temperature-induced cleavage because it harbors a reactive 2′ OH group in the ribose part of the ribonucleotides. At the temperatures at which hyperthermophiles thrive (> 80°C), the in vitro half-life of RNA molecules is measured in seconds. Since hyperthermophiles live happily at such high temperatures, they must have evolved efficient mechanisms to protect their macromolecules. High GC percentage, high intracellular concentration of branched and positively charged polyamines and RNA chaperons have all been found to be important for stabilization of tRNA. The presence of certain tRNA modifications is another important piece in the thermoresistance toolkit that hyperthermophiles have at their disposal. This is illustrated by larger diversity and abundance of tRNA modifications in hyperthermophiles as compared to mesophiles or psychrophiles. Modifications such as thiolations, methylations or even double methylations contribute to greater structural rigidity of tRNA and consequently they provide more resistance to thermal denaturation. The problematic 2′ OH group is often methylated in hyperthermophiles, which is highly effective against the heat-induced backbone breakage of tRNA. Finally, it seems that some thermophiles have evolved particular strategies for protection of their RNA. Recently, it was revealed that N^4-acetylcytidine (ac^4C) (see Figure 5.14 for chemical structure) is particularly abundant in RNA molecules (including tRNA and rRNA, see also section 5.1.4.3.1) from thermophilic *Thermococcales Euryarchaeota* but not in more distantly related *Methanocaldococcus jannaschii* or *Saccharolobus solfataricus*. The amounts of this modification were correlated with increase in temperature and its absence resulted in thermosensible phenotype, thus establishing this modification and the corresponding synthetic enzyme as thermoadaptation mechanism specific for *Thermococcales*.

As we have seen, the world of tRNA modifications is incredibly rich and, even though we are still far from complete understanding of diversity and function of tRNA modifications, it is clear that modified nucleosides are integral to tRNA functions at many levels influencing translation and tRNA structure and stability. In addition to these fundamental and well-documented functions, recent progress in the analyses of next-generation sequencing data unveiled numerous small noncoding RNAs with regulatory functions. Among those, the tRNA-derived fragments (tRFs) have attracted much attention because recent evidence shows that these are not merely degradation debris but true signaling molecules involved in many specific physiological and pathological processes in eukaryotes. Recent pioneering work reported the existence of regulatory tRFs in archaea (Figure 5.16) (see also sidebox), indicating that novel exciting avenues of tRNA-related research lie ahead.

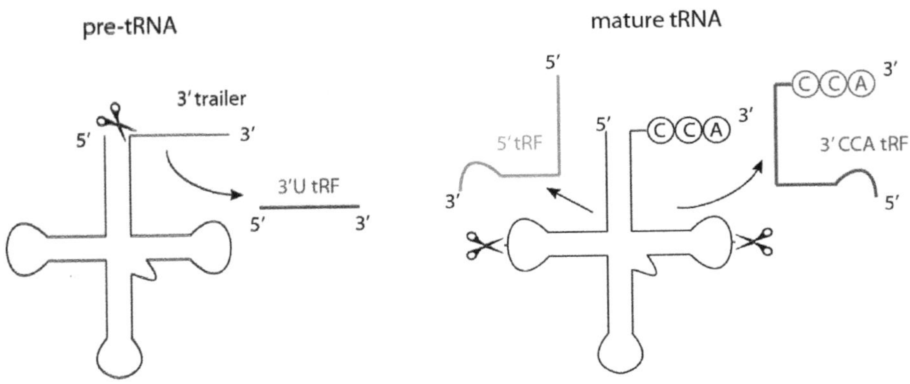

Figure 5.16. *tRNA-derived fragments in archaea (figure adapted from Clouet-d'Orval et al. 2018). For a color version of this figure, see www.iste.co.uk/clouet/archaea2.zip*

COMMENT ON FIGURE 5.16.– *Stable tRNA-derived fragments can be generated from precursor tRNAs (on the left) and from mature tRNAs (on the right). Endonucleolytic cleavage of the 3′ trailer sequence leads to the release of 3′U fragment. D-loop and T-loop of mature tRNAs can also be targeted by endonuclease(s) to yield 5′ or 3′ CCA fragments, respectively. Generation of stable tRNA-derived fragments (tRFs).*

Sidebox: tRNA-derived fragments in archaea: an unexplored universe

tRFs are small RNAs with size ranging from 14 to 30 nt. Several subtypes of tRFs exist based on their size and the site of the cleavage within the mature tRNA molecule or pre-tRNA. Despite substantial number of studies reporting the existence and the function of tRFs in eukaryotes, the details of specific mechanisms leading to their formation remains obscure. tRFs were only recently identified in archaea in several *Pyrobaculum sp.* and in *H. volcanii.* Remarkably, in the latter organism this phenomenon was detected for 11 out of 51 tRNAs and one of these was identified as major ribosome-binding RNA, which downregulated translation when cells were exposed to stressful conditions. This pioneering work indicates that vast universe of epigenetic tRNA-based regulation networks operates in archaea. The continuous progress in genetic manipulation of archaea, deep sequencing technologies and bioinformatics analyses should allow us to seriously tackle the investigation of diversity, biogenesis and function of tRFs in archaea in the coming years, with certainly many exciting discoveries to come.

5.2.5. *Aminoacylation of tRNAs*

To conclude this section, we will have a closer look at how a specific amino acid is attached to the corresponding tRNA. The enzymes that do this work with remarkable accuracy are called tRNA-synthetases (aaRSs). As expected, and given their universal requirement for protein synthesis, aaRSs show high level of evolutionary conservation but some eccentricities have been found in archaea and some of these noncanonical examples will be specifically highlighted in this section.

Before this, let us briefly review some general rules regarding the aminoacylation of tRNAs.

A total of 23 aaRSs have been described so far, one for each of the 20 proteinogenic amino acids (except for lysine, for which there are two) plus pyrrolysyl-tRNA synthetase (PylRS) and phosphoseryl-tRNA synthetase (SepRS) enzymes with a more restricted distribution that are only found in some bacterial and archaeal genomes (see below for more detail). These enzymes have to recognize, bind and transform three substrates: the correct amino acid (aa), the ATP that is needed to energetically fuel the reaction and the correct tRNA. The ATP is used to produce aa-AMP (this is called amino acid-adenylate) in the first step of the reaction. The aa-ATP intermediate is rich in energy and can therefore easily react with the 3′-OH group of the tRNA to form an ester bond and produce amino-acylated tRNA (aa-tRNA) (Figure 5.17).

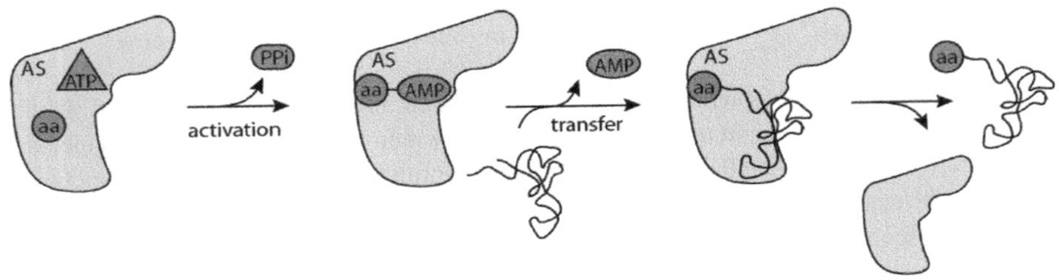

Figure 5.17. *Aminoacylation by aminoacyl tRNA synthetases (figure adapted from Yadavalli and Ibba 2012). For a color version of this figure, see www.iste.co.uk/clouet/archaea2.zip*

COMMENT ON FIGURE 5.17.– *The reaction proceeds in two main steps: first the aminoacyl tRNA synthetase (aaRS) activates amino acid (aa) via ATP hydrolysis in the aminoacylation active to form site (AS) an enzyme aminoacyl adenylate complex (aaRS·aa-AMP) with the release of inorganic pyrophosphate (PPi); second, the activated amino acid is transferred to the 3′-end of its cognate tRNA forming aminoacylated tRNA (aa-tRNA) followed by the dissociation of the enzyme and aminoacylated tRNA.*

aaRSs are divided into class I and class II enzymes, which differ in their active site architecture (defined by specific conserved sequence motifs), their mode of binding to the acceptor stem, and the position of the hydroxyl group they target for covalent attachment of the amino acid to tRNA. The class I enzymes recognize the minor groove of the acceptor stem and aminoacylate the 2′-OH of the ribose of A76, while the class II enzymes approach tRNA from the major groove and transfer amino acid to the 3′-OH group (Figure 5.18).

The accurate recognition of the appropriate tRNA-amino acid couple by specific aaRS is primordial for the survival of the cells. This is particularly challenging since aaRSs have to select the correct tRNA isoacceptor among the set of other tRNA molecules, which are all very similar in structure and chemical composition. In addition, aaRSs have to discriminate between a variety of similar proteinogenic and non-proteinogenic amino acids. In the case of tRNA, this difficult task is achieved by detection of subtle differences in shape of tRNA and with the help of identity determinants that are usually located in the anticodon stem or in the acceptor arm. Major known determinants are bases 35–37 next to the anticodon (which are also those that often carry modified nucleotides, see section 5.2.4) and the base 73 in the acceptor stem. It is also noteworthy that different identity determinants are used by different organisms and for each aaRS-tRNA-amino acid combination. Despite these challenges, aaRSs are remarkably accurate in selecting proper amino-acid and tRNA substrates and the misincorporation is estimated to occur only once in every 3,000 aminoacylation events in vivo.

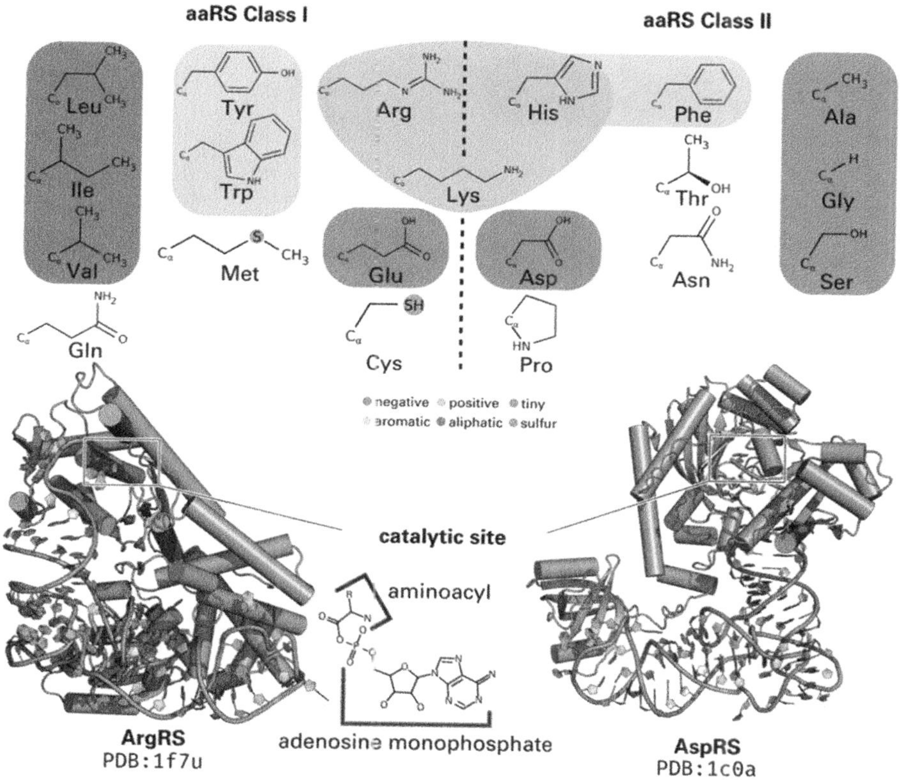

Figure 5.18. *Classification of aminoacyl-tRNA synthetases (figure reproduced from Kaiser et al. 2018). For a color version of this figure, see www.iste.co.uk/clouet/archaea2.zip*

COMMENT ON FIGURE 5.18.– *The 23 known aaRSs can be divided into two major classes based on the architecture of their catalytic sites (indicated with gray rectangle). These two classes also have a markedly different way in their modes of substrate binding. Class I aaRSs generally bind the minor groove of the tRNA acceptor stem and aminoacylate the 2'OH of the ribose of A76, while the class II aaRSs contact tRNA via the major groove and transfer amino acid to the 3'-OH. The ATP binding mode is also different: ATP is bound in its extended form by class I enzymes while the class II aaRSs bind a bent configuration where the gamma phosphate folds back over the adenine ring. Class I and class II aaRS recognize different sets of amino acids as substrates (shown on the top of the figure) with exception of lysine, which is mostly processed by class II aaRS encoded by* lysS *gene except in archaea where a class I aaRS encoded by* lysK *gene is responsible for lysine transfer to tRNA.*

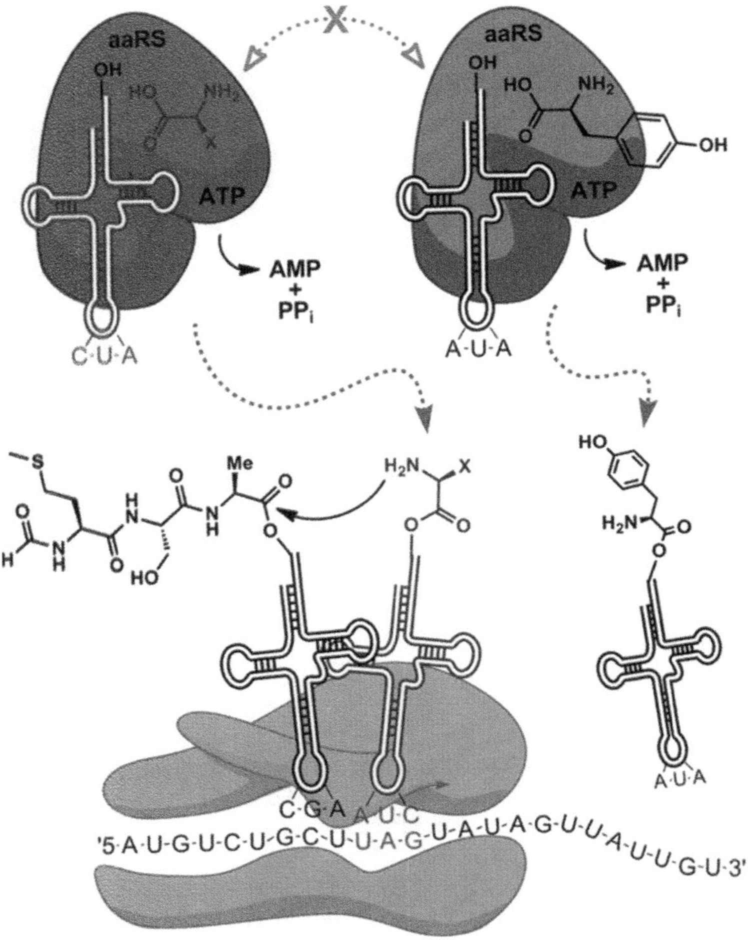

Figure 5.19. *Incorporation of non-canonical amino acids into proteins (figure reproduced from Young and Schultz 2010). For a color version of this figure, see www.iste.co.uk/clouet/archaea2.zip*

COMMENT ON FIGURE 5.19.– *Top part of the figure shows two aaRS-tRNA couples, one on the left is called orthogonal and the other is a natural, also called endogenous, couple (in this example TyrRS/tRNAAUA). The orthogonal pair must interact exclusively with each other and avoid cross-reactions with additional types of tRNAs and aaRS from a given organism (indicated by a gray two-sided arrow). In this example, the orthogonal aaRS aminoacylates the orthogonal tRNACUA (blue, red CUA) with an unnatural amino acid (blue X). In the lower part, translating ribosome is depicted. The orthogonal tRNA carries the unnatural amino acid and recognizes the UAG codon in the mRNA. Ribosome then incorporates the noncanonical amino acid into the growing polypeptide chain. In the past two decades, more than dozen of unnatural aaRS-tRNA couples were developed allowing to expand or rewrite the genetic code. Using these tools, more than 170 different noncanonical amino acids were site-specifically incorporated in proteins of bacteria and eukaryotes.*

Let us have a look now at the known peculiarities of the archaeal amino acyl-tRNA synthesis. One of the first that was discovered was the absence in some species of the full set of genes encoding all 23 aaRSs, indicating that archaea encode specific alternative pathways to synthetize certain aa-tRNAs. For instance, the gene encoding GlnRS is always absent in archaeal genomes and AsnRS is often missing. The studies of these peculiar cases led to the discovery of the first aaRS that violates the class rule (class I lysyl-tRNA synthetase instead of canonical class II enzyme) as well as new pathways for synthetizing aa-tRNAs from misacylated tRNAs (e.g. Asn-$tRNA^{Asn}$ from misacylated Asp-$tRNA^{Asn}$). More recently, a new archaeal pathway for selenocysteinyl-$tRNA^{Sec}$ involving a unique SepRS was discovered in some methanogens and it enabled recognizing that the equivalent (homologous) pathway also operates in eukaryotes.

The second noncanonical proteinogenic amino acid is pyrrolysine (Pyl), a derivative of lysine that was first discovered about 20 years ago during the study of *Methanosarcinae* archaea growth on methyl-amines. This revealed that one of the involved enzymes carried a functionally important pyrrolysin in its active site. Even more astonishing was the discovery that Pyl was encoded by UAG codon, which is one of the three STOP codons and, in principle, does not code an amino acid. To decode UAG, *Methanosarcinae* use a special $tRNA^{Pyl}$ that is amino acylated by a unique PylRS. The discovery of this particular $tRNA^{Pyl}$ and PylRS pair had a pivotal role in synthetic biology because it allowed us to rewrite the genetic code to allow for the incorporation of noncanonical amino acids into proteins (Figure 5.19).

The possibilities these tools offer are almost endless, for instance, one can design biosensors and biomarkers and study protein structure and function in live cells from bacteria to mammals.

5.3. Methodical toolboxes

Toolbox 5.1: biochemical isolation of (pre-)ribosomes

The structure and composition of native ribosomes and pre-ribosomes isolated from different model-organisms has been achieved over half a century. Nowadays, purifying native (pre-)ribosomes and identifying their associated components may feel a trivial everyday task, however, in some instances biochemical isolation remains challenging, especially in less biochemically characterized organisms.

Biochemical isolation of (pre-)ribosomal subunits is commonly based on differential or sedimentation (ultra)centrifugation and/or affinity-based purification. Classically, ribosomal subunits (and translationally active ribosome complexes) can be fractionated on sucrose/glycerol gradients according to their sedimentation coefficient using ultracentrifugation, and the resulting fractions can be further biochemically purified using additional gradient step or high-velocity centrifugation. Although pioneering identification of complex mixture of r-proteins was performed by 2D-PAGE and protein sequencing by Edman degradation, nowadays biological mass spectrometry strongly facilitates protein identification in complex mixture either from gel slice or directly from fractionated material. In contrast to r-proteins that are present in stoichiometric amounts within the very abundant ribosomal subunit, ribosome biogenesis factors are only associated with a very small fraction of immature ribosomal subunits that are typically underrepresented in such fractionated material.

To circumvent this problem and enriched for pre-ribosomal particles, affinity purification-based enrichment of defined pre-ribosomal intermediates coupled to downstream mass spectrometry identification of the co-purified proteins has been instrumental for the discovery of more than 200 eukaryotic ribosome biogenesis factors within a few years. To address this issue and enrich for pre-ribosomal particles, affinity purification of defined pre-ribosomal intermediates, combined with mass spectrometry to identify co-purified proteins, has been instrumental in discovering over 200 eukaryotic ribosome biogenesis factors within a few years. This approach relies on genetic engineering to fuse an affinity tag to a protein of interest or the use of a specific antibody that recognizes the bait protein, as well as optimized conditions for the native purification of transient interactors. These prerequisites, however, can be challenging to achieve depending on the organism studied.

The main challenge of all of these approaches is to maintain faithful conditions mimicking the native intra-cellular conditions in order to preserve transient and instable interactions.

Toolbox 5.2: RNA modification analysis

Mapping modified nucleotides on RNA sequences is notoriously laborious. Especially since three kinds of specific information are of interest: the chemical and positional assignments, and abundance/stoichiometry of these RNA modifications. Even though more than 100 different RNA modifications have been described, the chemical nature of some of these still escape our understanding (e.g. CN330 in archaeal/thermotoga 16S rRNA) or where only recently added to the RNA modification repertoire (e.g. NAD capping/glycosylated RNA).

Historically, RNA modification analysis was done by isolating a particular RNA species, labeling it radioactively, digesting it to single nucleotides and separating those using paper chromatography. This method required high amount of pure RNA as starting material and mostly could only identify a presence of a particular modification outside of any RNA sequence context. The RNA modification field has originally developed on the combination of ribonuclease digest of total/partly purified RNA combined with chromatography analysis. These approaches have enabled to identify the most abundant RNA modifications like 2′-O methylation and pseudouridylation. The advance of high-resolution and sensitive chromatography approach combined with the development of chemical standards, serving as references, have been key to obtain a complete RNA modification map of *S. cerevisiae* rRNA.

More recently, reverse transcriptase polymerase chain reaction (RT-PCR)-based methods and high-pressure liquid chromatography (HPLC) coupled to mass spectrometry (MS) were developed, allowing us to map a particular modification on RNA sequence. Mass spectrometry is currently the only technique that can comprehensively and directly characterize modifications in RNA molecules. This technique offers the possibility to analyze oligonucleotides resembling the native state of RNA molecules. Although separation of the oligonucleotides using MS compatible liquid chromatography remains challenging, progress has been made recently in automating interpretation of MS data by developing programs for database-matching or decoding of complicated patterns of oligonucleotide fragmentation. The main limitation of this technique is that fragmentation pattern obtained from analysis is searched against a reference database containing all possible combinations of RNA and known/possible chemical modifications but failed to identify directly chemical modification of unknown nature.

In parallel to the advanced mass spectrometry approaches, direct or indirect detection methods based on next-generation sequencing technologies were also developed. The existing techniques usually either use antibodies to enable the purification of RNAs containing the modification of interest from the total pool of RNAs. In parallel with advanced mass spectrometry approaches, direct and indirect detection methods based on next-generation sequencing technologies have also been developed. These techniques generally rely on two main strategies: one involves using antibodies to purify RNAs containing specific modifications from the total RNA pool. The other uses chemical derivatization to alter (or recode) RNA sequences at modified positions or to inhibit reverse transcription at these sites during cDNA library preparation. Together, these methods enable site-specific analysis of an increasing array of RNA modifications. The principal caveat of these approaches is that they are limited by a priori knowledge. In most cases, only one type of modification can be analyzed in each experiment and that specific antibodies and/or chemical derivation procedure for every known modification are not available. The detection of RNA modifications by direct RNA sequencing that takes advantage of Oxford nanopore technology, is presently being developed. In this approach, electrical current change signatures, which are specific to nucleotides (and derivatives), are recorded, while an RNA strand is transferred through a biological nanopore. Although this technology may potentially enable site-specific single molecule analysis of any virtual RNA modifications, it relies on a priori knowledge of RNA modification electrical current change signatures and the availability of reference to trained algorithms able to faithfully recognize and deconvolute these specific signatures.

Finally, the recent resolution revolution in cryogenic electron-microscopy (Cryo-EM) has already impacted RNA modification studies, whereby modified residues could be "directly" determined at atomic level within isolated ribosomal subunits. In the future, this technology may also further contribute to the de novo positioning of RNA modifications in native RNP, which can be biochemically purified and for which atomic resolution can be achieved.

5.4. References

Recommended literature for further reading

Archaea and phylogeny

Brochier-Armanet, C., Forterre, P., Gribaldo, S. (2011). Phylogeny and evolution of the Archaea: One hundred genomes later. *Curr Opin Microbiol.*, 14(3), 274–281.

Spang, A., Saw, J.H., Jørgensen, S.L., Zaremba-Niedzwiedzka, K., Martijn, J., Lind, A.E., van Eijk, R., Schleper, C., Guy, L., Ettema, T.J.G. (2015). Complex archaea that bridge the gap between prokaryotes and eukaryotes. *Nature*, 521, 173–179. doi: 10.1038/nature14447.

Tahon, G., Patricia Geesink, Ettema, T.J.G. (2021). Expanding archaeal diversity and phylogeny: Past, present, and future. *Annu. Rev. Microbiol.*, 75, 359–381. doi: 10.1146/annurev-micro-040921-050212.

Ribosome evolution

Bowman, J.C., Petrov, A.S., Frenkel-Pinter, M., Penev, P.I., Williams, L.D. (2020). Root of the tree: The significance, evolution, and origins of the ribosome. *Chemical Reviews*, 120(11), 4848–4878.

Petrov, A.S., Gulen, B., Norris, A.M., Kovacs, N.A., Bernier, C.R., Lanier, K.A., Fox, G.E., Harvey, S.C., Wartell, R M., Hud, N.V. et al. (2015). History of the ribosome and the origin of translation. *Proceedings of the National Academy of Sciences*, 112(50), 15396–15401.

Ribosome biogenesis in bacteria and eukarya

Baßler, J., and Hurt, E. (2019). Eukaryotic ribosome assembly. *Annu. Rev. Biochem.*, 88, 281–306. doi: 10.1146/annurev-biochem-013118-110817.

Davis, J.H., and Williamson, J.R. (2017). Structure and dynamics of bacterial ribosome biogenesis. *Philos. Trans. R. Soc. Lond. B Biol. Sci.*, 372(1716), 20160181. doi: 10.1098/rstb.2016.0181.

Henras, A.K., Plisson-Chastang, C., O'Donohue, M.-F., Chakraborty, A., Gleizes, P.-E. (2015). An overview of pre-ribosomal RNA processing in eukaryotes. *Wiley Interdisciplinary Reviews. RNA*, 6(2), 225–242.

Ribosome biogenesis in archaea

Ferreira-Cerca, S. (2017). Life and death of ribosomes in archaea. In *RNA Metabolism and Gene Expression in Archaea*, Clouet-d'Orval, B. (ed.). Springer, Cham. doi: 10.1007/978-3-319-65795-0_6.

Londei, P. and Ferreira-Cerca, S. (2021). Ribosome biogenesis in archaea. *Frontiers in Microbiology*, 12, 1476. doi: 10.3389/fmicb.2021.686977.

rRNA modifications

Lafontaine, D.L.J. and Tollervey, D. (1998). Birth of the snoRNPs: The evolution of the modification-guide snoRNAs. *Trends in Biochemical Sciences*, 23(10), 383–388.

Sharma, S. and Lafontaine, D.L.J. (2015). 'View from a bridge': A new perspective on eukaryotic rRNA base modification. *Trends in Biochemical Sciences*, 40(10), 560–575.

Sloan, K.E., Warda, A.S., Sharma, S., Entian, K.-D., Lafontaine, D.L.J., Bohnsack, M.T. (2016). Tuning the ribosome: The influence of rRNA modification on eukaryotic ribosome biogenesis and function. *RNA Biology*, 14(9), 1138–1152.

pre-tRNA processing

Bhatta, A. and Hillen, H.S. (2022). Structural and mechanistic basis of RNA processing by protein-only ribonuclease P enzymes. *Trends Biochem Sci.*, 47(11), 965–977. doi: 10.1016/j.tibs.2022.05.006.

Clouet-d'Orval, B., Batista, M., Bouvier, M., Quentin, Y., Fichant, G., Marchfelder, A., Maier, L.K. (2018). Insights into RNA-processing pathways and associated RNA-degrading enzymes in Archaea. *FEMS Microbiol Rev.*, 42(5), 579–613. doi: 10.1093/femsre/fuy016.

Hirata, A. (2019). Recent insights into the structure, function, and evolution of the RNA-splicing endonucleases. *Front Genet.*, 10, 103. doi: 10.3389/fgene.2019.00103.

Schiffer, S., Rösch, S., Marchfelder, A. (2002). Assigning a function to a conserved group of proteins: The tRNA 3'-processing enzymes. *EMBO J.*, 21(11), 2769–2777. doi: 10.1093/emboj/21.11.2769.

Wu, J., Niu, S., Tan, M., Huang, C., Li, M., Song, Y., Wang, Q., Chen, J., Shi, S., Lan, P., Lei, M. (2018). Cryo-EM structure of the human ribonuclease P holoenzyme. *Cell*, 175(5), 1393–1404.e11. doi: 10.1016/j.cell.2018.10.003.

tRNA modifications

El Yacoubi, B., Bailly, M., de Crécy-Lagard, V. (2012). Biosynthesis and function of posttranscriptional modifications of transfer RNAs. *Annu Rev Genet.*, 46, 69–95. doi: 10.1146/annurev-genet-110711-155641.

Phillips, G. and de Crécy-Lagard, V. (2011). Biosynthesis and function of tRNA modifications in Archaea. *Curr Opin Microbiol.*, 14(3), 335–341. doi: 10.1016/j.mib.2011.03.001.

Suzuki, T. (2021). The expanding world of tRNA modifications and their disease relevance. *Nat Rev Mol Cell Biol.*, 22(6), 375–392. doi: 10.1038/s41580-021-00342-0.

tRNA-derived fragments

Babski, J., Maier, L.K., Heyer, R., Jaschinski, K., Prasse, D., Jäger, D., Randau, L., Schmitz, R.A., Marchfelder, A., Soppa, J. (2014). Small regulatory RNAs in Archaea. *RNA Biol.*, 11(5), 484–493. doi: 10.4161/rna.28452.

Chen, Q., Zhang, X., Shi, J., Yan, M., Zhou, T. (2021). Origins and evolving functionalities of tRNA-derived small RNAs. *Trends Biochem Sci.*, 46(10), 790–804. doi: 10.1016/j.tibs.2021.05.001.

Gebetsberger, J., Zywicki, M., Künzi, A., Polacek, N. (2012). tRNA-derived fragments target the ribosome and function as regulatory non-coding RNA in *Haloferax volcanii*. *Archaea*, 2012, 260909. doi: 10.1155/2012/260909.

aminoacyl-tRNA synthetases

Rubio Gomez, M.A. and Ibba, M. (2020). Aminoacyl-tRNA synthetases. *RNA*, 26(8), 910–936. doi: 10.1261/rna.071720.119.

References (from the figure legends)

Armache, J.-P., Anger, A.M., Márquez, V., Franckenberg, S., Fröhlich, T., Villa, E., Berninghausen, O., Thomm, M., Arnold, G.J., Beckmann, R., Wilson, D.N. (2013). Promiscuous behaviour of archaeal ribosomal proteins: Implications for eukaryotic ribosome evolution. *Nucleic Acids Research*, 41(2), 1284–1293.

Ban, N., Beckmann, R., Cate, J.H.D., Dinman, J.D., Dragon, F., Ellis, S.R., Lafontaine, D.L.J., Lindahl, L., Liljas, A., Lipton, J.M. et al. (2014). A new system for naming ribosomal proteins. *Current Opinion in Structural Biology*, 24, 165–169.

Birikmen, M., Bohnsack, K.E., Tran, V., Somayaji, S., Bohnsack, M.T., Ebersberger, I. (2021). Tracing eukaryotic ribosome biogenesis factors into the archaeal domain sheds light on the evolution of functional complexity. *Frontiers in Microbiology*, 12, 2598. doi: 10.3389/fmicb.2021.739000.

Bowman, J.C., Petrov, A.S., Frenkel-Pinter, M., Penev, P.I., Williams, L.D. (2020). Root of the tree: The significance, evolution, and origins of the ribosome. *Chemical Reviews*, 120(11), 4848–4878.

Breuer, R., Gomes-Filho, J.-V., Randau, L. (2021). Conservation of archaeal C/D box sRNA-guided RNA modifications. *Frontiers in Microbiology*, 12, 496.

Brewer, T.E., Albertsen, M., Edwards, A., Kirkegaard, R.H., Rocha, E.P.C., Fierer, N. (2020). Unlinked rRNA genes are widespread among bacteria and archaea. *The ISME Journal*, 14(2), 597–608.

Brochier-Armanet, C., Forterre, P., Gribaldo, S. (2011). Phylogeny and evolution of the Archaea: One hundred genomes later. *Curr Opin Microbiol.*, 14(3), 274–281.

Coureux, P.-D., Lazennec-Schurdevin, C., Bourcier, S., Mechulam, Y., Schmitt, E. (2020). Cryo-EM study of an archaeal 30S initiation complex gives insights into evolution of translation initiation. *Communications Biology*, 3(1), 58–58.

Culver, G.M. (2003). Assembly of the 30S ribosomal subunit. *Biopolymers*, 68(2), 234–249.

Dennis, P.P., Tripp, V., Lui, L., Lowe, T., Randau, L. (2015). C/D box sRNA-guided 2′-O-methylation patterns of archaeal rRNA molecules. *BMC Genomics*, 16, 632.

Ebersberger, I., Simm, S., Leisegang, M.S., Schmitzberger, P., Mirus, O., von Haeseler, A., Bohnsack, M.T., Schleiff, E. (2014). The evolution of the ribosome biogenesis pathway from a yeast perspective. *Nucleic Acids Research*, 42(3), 1509–1523.

Ferreira-Cerca, S. (2017). Life and death of ribosomes in archaea. In *RNA Metabolism and Gene Expression in Archaea*, Clouet-d'Orval, B. (ed.). Springer, Cham. doi: 10.1007/978-3-319-65795-0_6.

Gerbi, S.A. (1986). The evolution of eukaryotic ribosomal DNA. *Biosystems*, 19(4), 247–258.

Gerbi, S.A. (1996). Expansion segments: Regions of variable size that interrupt the universal core secondary structure of ribosomal RNA. *Ribosomal RNA Structure, Evolution, Processing, and Function in Protein Biosynthesis*, 1996, 71–87.

Grosjean, H., Gaspin, C., Marck, C., Decatur, W.A., de Crécy-Lagard, V. (2008). RNomics and Modomics in the halophilic archaea *Haloferax volcanii*: Identification of RNA modification genes. *BMC Genomics*, 9, 470. doi: 10.1186/1471-2164-9-470.

Grosjean, H., Breton, M., Sirand-Pugnet, P., Tardy, F., Thiaucourt, F., Citti, C., Barré, A., Yoshizawa, S., Fourmy, D., de Crécy-Lagard, V. et al. (2014). Predicting the minimal translation apparatus: Lessons from the reductive evolution of mollicutes. *PLoS Genetics*, 10(5), e1004363.

Grünberger, F., Knüppel, R., Jüttner, M., Fenk, M., Borst, A., Reichelt, R., Hausner, W., Soppa, J., Ferreira-Cerca, S., Grohmann, D. (2020). Exploring prokaryotic transcription, operon structures, rRNA maturation and modifications using Nanopore-based native RNA sequencing. *bioRxiv*. doi: 10.1101/2019.12.18.880849.

Hadjiolov, A.A. (1985). *The Nucleolus and Ribosome Biogenesis*. Springer, Vienna.

Hage, A.E. and Tollervey, D. (2004). A surfeit of factors: Why is ribosome assembly so much more complicated in Eukaryotes than bacteria? *RNA Biology*, 1(1), 9–14.

Henras, A.K., Plisson-Chastang, C., O'Donohue, M.-F., Chakraborty, A., Gleizes, P.-E. (2015). An overview of pre-ribosomal RNA processing in eukaryotes. *Wiley Interdisciplinary Reviews: RNA*, 6(2), 225–242.

Jüttner, M. and Ferreira-Cerca, S. (2022). Looking through the lens of the ribosome biogenesis evolutionary history: Possible implications for archaeal phylogeny and eukaryogenesis. *Molecular Biology and Evolution*, 39(4), msac054.

Jüttner, M., Weiß, M., Ostheimer, N., Reglin, C., Kern, M., Knüppel, R., Ferreira-Cerca, S. (2020). A versatile cis-acting element reporter system to study the function, maturation and stability of ribosomal RNA mutants in archaea. *Nucleic Acids Research*, 48(4), 2073–2090.

Klappenbach, J.A., Saxman, P.R., Cole, J.R., Schmidt, T.M. (2001). rrndb: The ribosomal RNA operon copy number database. *Nucleic Acids Research*, 29(1), 181–184.

Knüppel, R., Christensen, R.H., Gray, F.C., Esser, D., Strauß, D., Medenbach, J., Siebers, B., MacNeill, S.A., LaRonde, N., Ferreira-Cerca, S. (2018). Insights into the evolutionary conserved regulation of Rio ATPase activity. *Nucleic Acids Research*, 46(3), 1441–1456.

Knüppel, R., Trahan, C., Kern, M., Wagner, A., Grünberger, F., Hausner, W., Quax, T.E.F., Albers, S.-V., Oeffinger, M., Ferreira-Cerca, S. (2021). Insights into synthesis and function of KsgA/Dim1-dependent rRNA modifications in archaea. *Nucleic Acids Research*, 49(3), 1662–1687.

Kovacs, N.A., Penev, P.I., Venapally, A., Petrov, A.S., Williams, L.D. (2018). Circular permutation obscures universality of a ribosomal protein. *Journal of Molecular Evolution*, 86(8), 581–592.

Krogh, N., Jansson, M.D., Häfner, S.J., Tehler, D., Birkedal, U., Christensen-Dalsgaard, M., Lund, A.H., Nielsen, H. (2016). Profiling of 2'-O-Me in human rRNA reveals a subset of fractionally modified positions and provides evidence for ribosome heterogeneity. *Nucleic Acids Research*, 44(16), 7884–7895.

Lafontaine, D.L.J. and Tollervey, D. (1998). Birth of the snoRNPs: The evolution of the modification-guide snoRNAs. *Trends in Biochemical Sciences*, 23(10), 383–388.

Lecompte, O., Ripp, R., Thierry, J., Moras, D., Poch, O. (2002). Comparative analysis of ribosomal proteins in complete genomes: An example of reductive evolution at the domain scale. *Nucleic Acids Research*, 30(24), 5382–5390.

Leffers, H., Kjems, J., Østergaard, L., Larsen, N., Garrett, R.A. (1987). Evolutionary relationships amongst archaebacteria: A comparative study of 23 S ribosomal RNAs of a sulphur-dependent extreme thermophile, an extreme halophile and a thermophilic methanogen. *Journal of Molecular Biology*, 195(1), 43–61.

Londei, P. and Ferreira-Cerca, S. (2021). Ribosome biogenesis in archaea. *Frontiers in Microbiology*, 12, 1476. doi: 10.3389/fmicb.2021.686977.

Londei, P., Teixidò, J., Acca, M., Cammarano, P., Amils, R. (1986). Total reconstitution of active large ribosomal subunits of the thermoacidophilic archaebacterium *Sulfolobus solfataricus*. *Nucleic Acids Research*, 14(5), 2269–2285.

Mangiarotti, G. and Chiaberge, S. (1997). Reconstitution of functional eukaryotic ribosomes from *Dictyostelium discoideum* ribosomal proteins and RNA. *Journal of Biological Chemistry*, 272(32), 19682–19687.

Márquez, V., Fröhlich, T., Armache, J.-P., Sohmen, D., Dönhöfer, A., Mikolajka, A., Berninghausen, O., Thomm, M., Beckmann, R., Arnold, G.J. et al. (2011). Proteomic characterization of archaeal ribosomes reveals the presence of novel archaeal-specific ribosomal proteins. *Journal of Molecular Biology*, 405(5), 1215–1232.

Nakao, A., Yoshihama, M., Kenmochi, N. (2004). RPG: The ribosomal protein gene database. *Nucleic Acids Research*, 32, D168–D170.

Nierhaus, K.H. (1991). The assembly of prokaryotic ribosomes. *Biochimie*, 73(6), 739–755.

Nierhaus, K.H. and Lafontaine, D.L. (2004). Ribosome assembly. In *Protein Synthesis and Ribosome Structure*. Wiley-VCH, Weinheim. doi: 10.1002/3527603433.ch3.

Noller, H.F., Kop, J., Wheaton, V., Brosius, J., Gutell, R.R., Kopylov, A.M., Dohme, F., Herr, W., Stahl, D.A., Gupta, R. et al. (1981). Secondary structure model for 23S ribosomal RNA. *Nucleic Acids Research*, 9(22), 6157–6189.

Nürenberg-Goloub, E., Kratzat, H., Heinemann, H., Heuer, A., Kötter, P., Berninghausen, O., Becker, T., Tampé, R., Beckmann, R. (2020). Molecular analysis of the ribosome recycling factor ABCE1 bound to the 30S post-splitting complex. *The EMBO Journal*, 39(9), e103788.

Parker, M.S., Sallee, F.R., Park, E.A., Parker, S.L. (2015). Homoiterons and expansion in ribosomal RNAs. *FEBS Open Bio*, 5(1), 864–876.

Penev, P.I., Fakhretaha-Aval, S., Patel, V.J., Cannone, J.J., Gutell, R.R., Petrov, A.S., Williams, L.D., Glass, J.B. (2020). Supersized ribosomal RNA expansion segments in Asgard archaea. *Genome Biology and Evolution*, 12(10), 1694–1710.

Petrov, A.S., Gulen, B., Norris, A.M., Kovacs, N.A., Bernier, C.R., Lanier, K.A., Fox, G.E., Harvey, S.C., Wartell, R.M., Hud, N.V. et al. (2015). History of the ribosome and the origin of translation. *Proceedings of the National Academy of Sciences*, 112(50), 15396–15401.

Sanchez, M.E., Urena, D., Amils, R., Londei, P. (1990). In vitro reassembly of active large ribosomal subunits of the halophilic archaebacterium *Haloferax mediterranei*. *Biochemistry*, 29(39), 9256–9261.

Sanchez, M.E., Londei, P., Amils, R. (1996). Total reconstitution of active small ribosomal subunits of the extreme halophilic archaeon *Haloferax mediterranei*. *Biochimica et Biophysica Acta (BBA) – Protein Structure and Molecular Enzymology*, 1292(1), 140–144.

Sas-Chen, A., Thomas, J.M., Matzov, D., Taoka, M., Nance, K.D., Nir, R., Bryson, K.M., Shachar, R., Liman, G.L.S., Burkhart, B.W. et al. (2020). Dynamic RNA acetylation revealed by quantitative cross-evolutionary mapping. *Nature*, 583(7817), 638–643.

Schwarz, T.S., Berkemer, S.J., Bernhart, S.H., Weiß, M., Ferreira-Cerca, S., Stadler, P.F., Marchfelder, A. (2020). Splicing endonuclease is an important player in rRNA and tRNA maturation in archaea. *Frontiers in Microbiology*, 11, 594838. doi: 10.3389/fmicb.2020.594838.

Sharma, S. and Lafontaine, D.L.J. (2015). 'View from a bridge': A new perspective on eukaryotic rRNA base modification. *Trends in Biochemical Sciences*, 40(10), 560–575.

Sloan, K.E., Warda, A.S., Sharma, S., Entian, K.-D., Lafontaine, D.L.J., Bohnsack, M.T. (2016). Tuning the ribosome: The influence of rRNA modification on eukaryotic ribosome biogenesis and function. *RNA Biology*, 14(9), 1138–1152.

Stoddard, S.F., Smith, B.J., Hein, R., Roller, B.R.K., Schmidt, T.M. (2015). rrnDB: Improved tools for interpreting rRNA gene abundance in bacteria and archaea and a new foundation for future development. *Nucleic Acids Research*, 43(D1), D593.

Taoka, M., Nobe, Y., Yamaki, Y., Sato, K., Ishikawa, H., Izumikawa, K., Yamauchi, Y., Hirota, K., Nakayama, H., Takahashi, N. et al. (2018). Landscape of the complete RNA chemical modifications in the human 80S ribosome. *Nucleic Acids Research*, 46(18), 9289–9298.

Thomson, E., Ferreira-Cerca, S., Hurt, E. (2013). Eukaryotic ribosome biogenesis at a glance. *Journal of Cell Science*, 126(21), 4815.

Tu, J. and Zillig, W. (1982). Organization of rRNA structural genes in the archaebacterium *Thermoplasma acidophilum*. *Nucleic Acids Research*, 10(22), 7231–7245.

Warner, J.R. (1999). The economics of ribosome biosynthesis in yeast. *Trends in Biochemical Sciences*, 24(11), 437–440.

Watkins, N.J. and Bohnsack, M.T. (2012). The box C/D and H/ACA snoRNPs: Key players in the modification, processing and the dynamic folding of ribosomal RNA. *WIREs RNA*, 3(3), 397–414.

Woese, C.R. and Fox, G.E. (1977). Phylogenetic structure of the prokaryotic domain: The primary kingdoms. *Proceedings of the National Academy of Sciences*, 74(11), 5088–5090.

Woese, C.R., Kandler, O., Wheelis, M.L. (1990). Towards a natural system of organisms: Proposal for the domains Archaea, Bacteria, and Eucarya. *Proceedings of the National Academy of Sciences of the United States of America*, 87(12), 4576–4579.

Woolford, J.L. and Baserga, S.J. (2013). Ribosome biogenesis in the yeast *Saccharomyces cerevisiae*. *Genetics*, 195(3), 643–681.

Young, T.S. and Schultz, P.G. (2010). Beyond the canonical 20 amino acids: Expanding the genetic lexicon. *Journal of Biological Chemistry*, 285(15), 11039–11044.

Yutin, N., Puigbò, P., Koonin, E.V., Wolf, Y.I. (2012). Phylogenomics of prokaryotic ribosomal proteins. *PLoS ONE*, 7(5), e36972.

6

The Diversity and Function of Noncoding RNAs in Archaea

Hubert F. BECKER[1,2] **and Christine GASPIN**[3]

[1]*Laboratoire d'optique et biosciences, Inserm, CNRS, Institut Polytechnique de Paris, Palaiseau, France*

[2]*Sorbonne Université, Paris, France*

[3]*MIAT, INRAE, Université de Toulouse, Castanet-Tolosan, France*

6.1. Introduction

Noncoding RNAs (ncRNAs) are present in all three domains of life and are involved as major players in the control of cellular base functions, organism development and stress response. They form a class of transcripts (RNAs) that are not translated into proteins, but function directly as RNAs. The term "small RNA" (sRNA) is also found in the literature to designate small ncRNAs in archaea.

The best-known ncRNAs, shared by all three domains of life, are ribosomal RNA (rRNA), transfer RNA (tRNA), signal recognition particle RNA (7S or SRP RNA) and ribonuclease P RNA (RNaseP RNA). rRNA is a component of the ribosome and contributes to the translation of messenger RNAs (mRNAs) into proteins. The ribosome comprises large and small subunits. The large subunit integrates 23S rRNA and 5S rRNA, and forms the catalytic core of the ribosome, together with the peptidyl transferase center, which synthesizes peptide bonds between amino acids in the peptide chain being synthesized. The small subunit, composed of 16S RNA, is responsible for reading mRNA codons. Linked to the translation machinery, each tRNAs transports one of the 20 amino acids that will help decode the mRNA, to produce the chain of amino acids making up the protein.

Biology of Archaea 2,
coordinated by Béatrice CLOUET-D'ORVAL, Bruno FRANZETTI and Philippe OGER.

SRP RNA is also a key player in the translation machinery. It contributes to the recognition of a signal peptide (or addressing sequence), which indicates to the cell the compartment to which to address the protein sequences generated. Lastly, RNase P RNA contributes to tRNA maturation.

Other ncRNAs exist in archaea, such as regulatory RNAs, introns, riboswitches, modification guide RNAs and CRISPR ncRNAs (clustered regularly interspaced short palindromic repeats, CRISPR ncRNA (crRNAs)), which have been demonstrated in bacteria as defense systems against foreign elements. Riboswitches form a large subclass of ncRNAs, which are universally shared. They correspond to sequences of regulatory regions, which are located in the 5′ untranslated part of mRNAs. Riboswitches have mainly been characterized in bacteria, but remain largely unexplored in archaea. Remarkably, archaea share two families of ncRNAs with eukaryotes, box C/D and box H/ACA ncRNAs, involved in 2'O-methylation and pseudo-uridylation. These ncRNAs act by pairing at targeted positions on the rRNAs and tRNAs to be modified. In archaea, in silico identification of members of the above-mentioned families is now possible at the nucleic acid sequence level of complete genomes.

More recently, high-throughput transcriptome sequencing technologies (RNA-seq) have considerably accelerated the discovery of ncRNAs, which were unexpectedly numerous and diverse in all three domains of life. In archaea, it is only recently that large-scale analyses of noncoding transcripts have made it possible to identify several new major ncRNA subclasses (Figure 6.1), including cis antisense ncRNAs (asRNA), sense-overlapping ncRNAs (sRNAs), trans-regulatory ncRNAs (regRNAs), long ncRNAs (lncRNAs) and tRNA-derived small RNAs (tsRNAs).

In this chapter, we review computational and experimental methods for identifying ncRNAs, as well as describe the main sub-classes of ncRNAs known in archaea, distinguishing them according to their level of characterization, their mode of action, their role, their size or their involvement in a specific process.

6.2. Noncoding RNA diversity

ncRNAs are responsible for many essential cellular processes. The best-known of these, rRNA and tRNA, are described in detail in Chapter 5 of this book, which deals with ribosome biogenesis and translation. rRNAs and tRNAs respectively catalyze the translation of mRNA into protein, and the decoding of mRNA with amino acids for peptide chain synthesis, through specific tertiary interactions between RNA molecules.

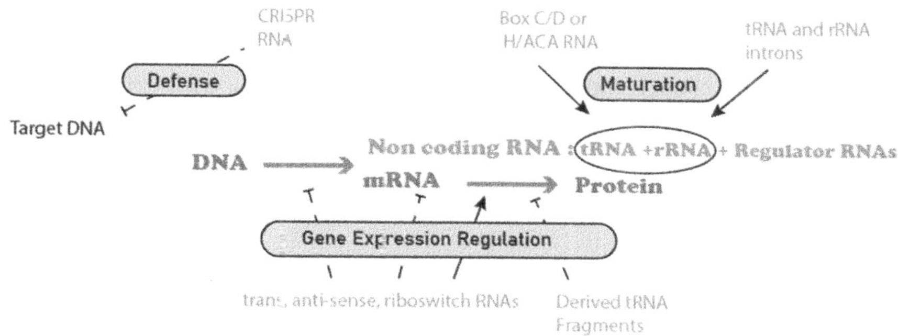

Figure 6.1. *Types of noncoding RNA-related regulatory mechanisms in archaea. Positive and negative regulation of transcription and translation steps are represented by solid or dotted black arrows, respectively. Noncoding RNAs are in orange, and transcription and translation steps are represented by gray arrows. For a color version of this figure, see www.iste.co.uk/clouet/archaea2.zip*

The other ncRNAs present in archaea are often referred to as "small RNAs", because of their size of less than 500 nucleotides. Figure 6.1 shows a schematic representation of the ncRNA interaction network in archaea, during the processes of RNA maturation, gene expression regulation and defense. These include guide RNAs (box C/D and H/ACA RNAs) involved in RNA modification, cis-regulatory and trans-regulatory RNAs, tRNA-derived fragments and CRISPR RNAs. The vast majority of these RNAs have been identified in recent years thanks to "RNomics" technologies, combined with bioinformatics analysis. These ncRNAs have been shown to play a key role in the regulation of gene expression. Since archaea share a wide diversity of metabolic pathways with eukaryotes and bacteria, they logically share a series of regulatory RNAs with expression levels specific to each. These ncRNAs usually respond to environmental stimuli and adaptations to extreme conditions of temperature or osmolarity in order to regulate various biological processes, such as enzymatic activities, metabolite requirements, stress response, or the definition of cell morphology.

6.2.1. *ncRNA modification guides*

Box C/D and H/ACA ncRNAs, first identified in eukaryotes with the name snoRNA (for Small Nucleolar RNA) due to their localization in the nucleoli, are also present in archaea (Gaspin et al. 2000; Omer 2000), although the latter lack nucleoli. Box C/D and H/ACA ncRNAs represent two abundant classes of ncRNAs involved in 2′-O methylation and pseudouridylation of rRNAs and tRNAs, respectively. In archaea, their number can vary from less than a dozen in *Metanococcus jannashi* to several dozen in *Thermococcales*, and they are generally localized in intergenic regions, sometimes partially overlapping ends of open reading frames (ORF) (Gaspin et al. 2000; Omer 2000). Absent in bacteria, ncRNAs with C/D or H/ACA boxes probably preexisted the archaea/eukaryote separation (Hüttenhofer and Schattner 2006).

6.2.1.1. *Box C/D ncRNAs*

Box C/D ncRNAs guide the 2′-O-methylation of riboses. They are characterized by the presence of conserved sequence segments, called C and C′ boxes (RUGAUGA) and D and D′ boxes (CUGA), which are essential for ncRNA stability, biogenesis and operation (for a review, see Omer et al. 2003). Their size is, on average, much shorter (45–60 nt) than in eukaryotes (60 to over 100 nt) (Gaspin et al. 2000; Omer 2000). The predicted number of potentially 2′-O-methylated residues could be correlated with the optimal growth temperature of archaea (Omer 2000; Dennis et al. 2001), but the number of species analyzed does not allow us to validate this hypothesis. Boxes C and D are located close to the 5′ (box C) and 3′ (box D) at the end of the ncRNA (Figure 6.2(a)). In archaea, they are more conserved than in eukaryotes (Gaspin et al. 2000). Box C/D ncRNAs contain one or two guide sequences that are complementary to the sequence of their target RNA, over a length of around ten nucleotides. The modified position on the target is matched to the fifth nucleotide upstream of the ncRNA's D box or D′ box (Figure 6.2(a)). The nature of the sequence preceding the D and D′ boxes is therefore the determinant of the modified site. Box C/D ncRNAs predominantly target rRNA, but several studies have also demonstrated their role in tRNA nucleotide modification (Omer 2000; Clouet-d'Orval et al. 2001; Dennis et al. 2001; Renalier et al. 2005). Remarkably in archaea, the mature product of the tRNA Trp intron is a box C/D ncRNA involved in the 2′-O methylation of two positions of the tRNA Trp that carries it (Clouet-d'Orval et al. 2001). One of these two modifications is essential for tRNA maturation by recognition of a characteristic BHB (bulge–helix–bulge) structure. As in plants, examples of dicistronic tRNA-ncRNA organization with C/D boxes have been demonstrated in *Nanoarchaeum equitans* and *Sulfolobus acidocaldarius*. Another example is the presence of a box C/D ncRNA in the primary transcript of the rRNA operon (Tang 2002). Certain positions of modified rRNA nucleotides are conserved in eukaryotes (Gaspin et al. 2000), revealing the importance of these modifications. Unlike eukaryotic box C/D snoRNAs, which predominantly target a single position, the vast majority of archaeal box C/D ncRNAs target two positions, often very close in sequence or structure. This feature suggests a contribution to rRNA folding (Gaspin et al. 2000; Omer 2000; for a review, see Bachellerie et al. 2002). Box C/D ncRNAs form a structural motif at the C/D and C′/D′ boxes known as the K-turn/K-loop (Figure 6.2(a)) (Nolivos 2005) and, at their ends, a characteristically short terminal helix compared with eukaryotes. Their circularization by ligation of their 5′ and 3′ ends seems to be a necessary component for their stability, so as to compensate for the lack of a stable terminal helix (Starostina et al. 2004; Becker et al. 2017). The associated ribonucleoprotein complex is composed of three essential proteins (Figure 6.2(a)) (for a review, see Reichow et al. 2007). Fibrillarin (also known as Nop1 or Fib1) supports catalytic activity. The Nop5 protein is more specifically associated with the C/D and C′/D′ boxes. The L7Ae protein binds to the K-turn and K-loop motifs formed by the C′/D′ box pair (Rozhdestvensky 2003; Charron et al. 2004).

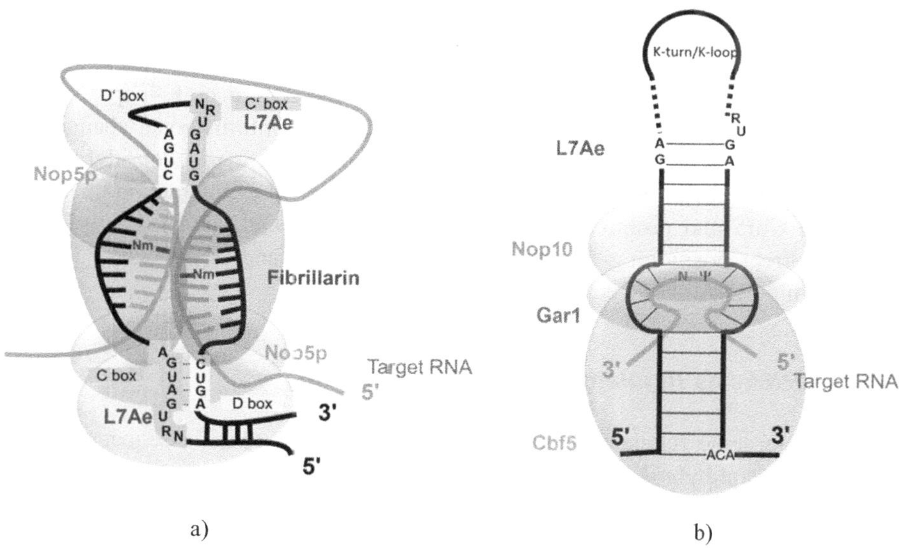

Figure 6.2. *Schematic representation of ribonucleic complexes involving (a) box C/D and (b) box H/ACA ncRNAs. The ncRNA sequence is shown in black. C/C′ boxes (UGAUGA) are shown with a blue highlight, and D/D′ boxes (CUGA) or H/ACA boxes are shown with a yellow highlight. The rRNA sequence is shown in gray. Modified positions are shown in red. Proteins interacting in the complex are indicated on the interaction region. For a color version of this figure, see www.iste.co.uk/clouet/archaea2.zip*

6.2.1.2. *ncRNAs with H/ACA boxes*

Box H/ACA ncRNAs are characterized by the presence of conserved segments called H (ANANNA) and ACA (ANA) boxes (for a review, see Reichow et al. 2007). In archaea, these ncRNAs adopt a secondary structure of one, two or three stem-loops, whereas two (sometimes one) stem-loops are typical in eukaryotes. In archaea, each stem-loop contains a K-turn or K-loop motif (Figure 6.2(b)). When the ncRNA contains two or three stem-loops, the H-box is located in the short region between them. In archaea, it may show a shorter consensus motif (ANA), similar to the ACA box which is located at the 3′ end of the ncRNA. Each stem-loop features an inner loop, also known as a "pseudouridylation pocket", whose unpaired segments are complementary to the target RNA around the modified position. The two complementary segments on rRNA are located, respectively, two nucleotides downstream and directly upstream of the targeted U residue (Figure 6.2(b)). On box H/ACA ncRNAs, the distance of the H or ACA box from the helix closing the inner loop is always between 13 and 15 nucleotides. The ribonucleic complex associated with box H/ACA ncRNA also comprises proteins including Nop10, L7Ae, Cbf5 and Gar1 (for a review, see Watkins and Bohnsack 2012). Cbf5 binds to ncRNA via H/ACA boxes (Baker 2005). L7Ae recognizes the K-turn or K-loop structure present in the upper end of the stem-loop and induces a conformational change that positions the targeted "U" in the active site of Cbf5, where pseudouridylation is performed (Rozhdestvensky 2003). Nop10 organizes the interaction between L7Ae and Cbf5 and promotes target binding (Hamma et al. 2005). Lastly, Gar1 ensures correct positioning of the modified U position and is necessary for the rerelease of ncRNA.

6.2.2. *ncRNAs as regulators of gene expression*

Each intergenic regulator ncRNA (regRNA) targets a specific sequence of an mRNA by partial hybridization with it, usually resulting in inhibition of translation. Only a few ncRNA targets have been described to date, with different modes of interaction between ncRNAs and mRNAs (Babski et al. 2014; Gelsinger and DiRuggiero 2018a) (Figure 6.3). The examples of ncRNA interaction mechanisms shown in Figure 6.3(b) are taken from several studies of ncRNAs in archaea: of the 18 ncRNAs identified in *M. mazei*, the ncRNA sRNA162 can form either a trans interaction with the internal ribosome entry region (RBS) of the MM2440-MM2441 mRNA, or a cis interaction with the 5′ region of MM2242, resulting in a block in translation and/or mRNA degradation (Jäger et al. 2012).

As observed for other regRNAs, it would appear that regRNAs are capable of associating with multiple target sequences of the same or multiple mRNAs. In the case of mRNAs lacking a 5′-UTR region, it has been observed that the 3′-UTR region may also be involved in regulating translation or transcript stabilization, through interaction with a regRNA (Gomes-Filho et al. 2018). Another mode of regulation has been described in the archaea *S. acidocaldarius*, in which ncRNA interacts directly with the mRNA coding sequence, playing the role of antisense RNA (Orell et al. 2018).

A second pathway for regulating gene expression, known as cis-regulation, is provided by riboswitches, which are highly structured 5′-UTR regions of mRNA that are sensitive to temperature or metabolite binding (Figure 6.3). The structural rearrangement induced on the mRNA will modulate the expression of genes encoded on the corresponding mRNA, either by allowing the elongation step of transcription to continue, or by causing transcriptional arrest. Some archaeal riboswitches can also restrict or promote access to the translation initiation site, representing another means of regulation. In the archaea *T. kodakarensis*, a fluorine-sensitive riboswitch has been described as a native and active regulator of the expression of genes implicated in fluorine detoxification (Speed et al. 2018).

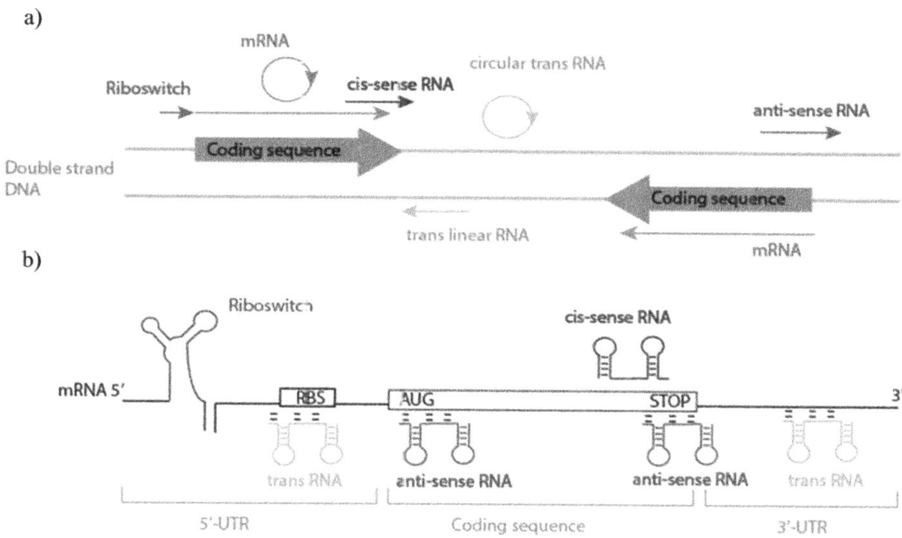

Figure 6.3. *Noncoding RNA diversity in archaea and mechanisms of interaction. (a) Different types of noncoding RNA are shown, originating from transcription initiation sites, intergenic regions, and sense-oriented or antisense-oriented genes; (b) schematic diagram of the mechanisms of interaction between noncoding RNA and the coding, 5'-UTR and 3'-UTR regions of an mRNA. For a color version of this figure, see www.iste.co.uk/clouet/archaea2.zip*

6.2.3. *CRISPR ncRNA*

crRNAs are ncRNAs in bacteria and archaea that are essential for defense against the invasion of foreign DNA, such as viral genomes, plasmid conjugates or other mobile genetic elements (Figure 6.4). This immunity is directed against any new foreign DNA that is identical or close in sequence to previously integrated

DNA. In the genome, foreign DNA is integrated in the form of short sequences, also known as spacers, which are surrounded by repeated sequences and grouped together in the same region, leading to an organization known as CRISPR. The CRISPR locus will enable the production of crRNAs which, in association with the endonucleases (Cas) of the CRISPR system, will degrade the foreign RNA in case of a new invasion (Makarova et al. 2015).

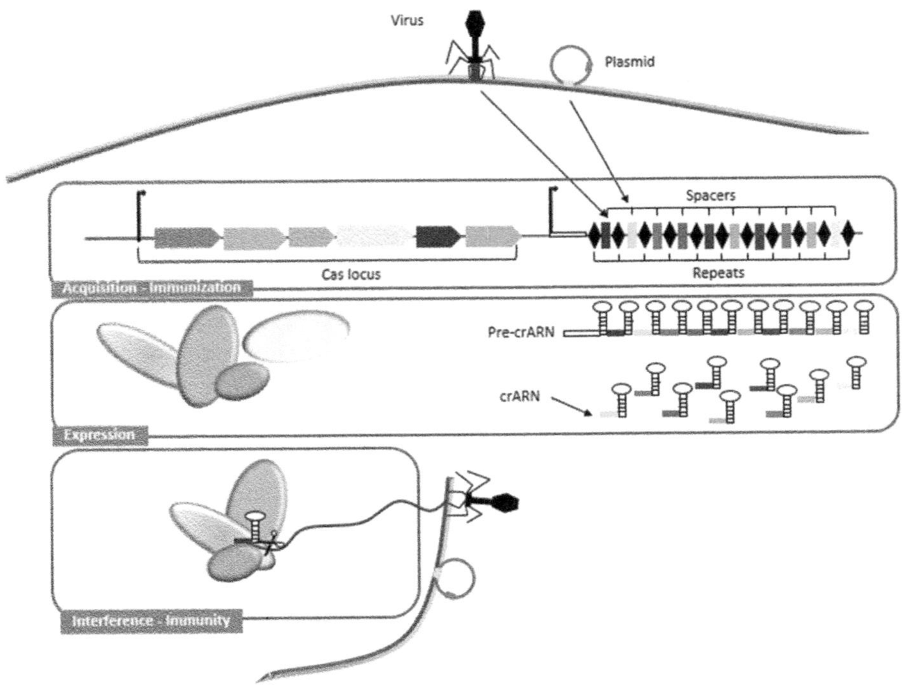

Figure 6.4. *CRISPR system. In three steps, immunization of the organism is provided by the integration of short gene sequences from phages or plasmids (acquisition–immunization), and immunity is provided by the expression of crRNAs as guide RNAs (expression), which associate with the Cas complex to destroy any pathogen intrusion (interference–immunity). For a color version of this figure, see www.iste.co.uk/clouet/archaea2.zip*

The exact molecular mechanism by which crRNAs will regulate gene expression in archaea is not known. However, their existence has been revealed in several archaea species by RNomics experiments or co-immunoprecipitation with different bait proteins (Marchfelder et al. 2012). Bioinformatic analyses of genomes have revealed the widespread presence of CRISPR loci, suggesting that the CRISPR autoimmunity mechanism is present in 90% of archaea, compared with 40% in bacterial genomes, with a predominance in thermophilic species compared with mesophiles and psychrophiles. Several studies have been carried out on archaeal CRISPR/Cas systems in order to determine the organization of the CRISPR/Cas complex, obtain crystallographic structures of Cas proteins, identify the system's target sequences and solve the mechanism of hydrolysis (Marchfelder et al. 2012). All of these data will ultimately help define the extent and importance of CRISPR/Cas systems in cellular processes.

6.2.4. *Small RNAs derived from tRNAs (tsRNAs)*

Initially discovered in eukaryotic cells, tsRNAs have been identified in several archaeal species. Fragmentation of tRNAs is not a general cellular process, but is only observed for certain tRNAs under certain conditions of cellular stress. It is now accepted that tsRNAs are no longer simply products of degradation or aberrant maturation of tRNAs, but play functional regulatory roles in numerous biological processes, such as cell proliferation, protein biosynthesis or interference mechanisms via microRNAs (miRNAs) and interfering RNAs (siRNAs – silencing) (for a review, see Gebetsberger and Polacek 2013). Different categories of tsRNA have clearly been identified in the three domains of life. For example, (i) tRNA halves, tiRNAs (tRNA-derived stress-induced RNA, 28–36 nucleotides) derived from specific cuts in the anticodon loop of mature tRNAs; (ii) smaller tRNA fragments, tRF (tRNA-derived fragments, 14–30 nucleotides) derived from non-randomized cuts, outside the anticodon loop, in pre-rRNA or mature tRNA, forming the sub-categories tRF-5, tRF-3 and tRF-1. The tRF-5, tRF-3 fragments are generated from the 5′ and 3′ ends of mature tRNAs, while the tRF-1 fragment is derived from the 3′-UTR end of the pre-tRNA (Figure 6.5) (Gebetsberger and Polacek 2013; Xie et al. 2020).

In archaea, tsRNAs have been described for several species of *Pyrobaculum* (Bernick et al. 2012a) and *H. volcanii*, with the identification of tsRNAs for 11 out of 51 tRNAs in the latter case, by RNA-seq and northern blot (Heyer et al. 2012; Gebetsberger et al. 2017). In other studies, many tRNA fragments have been shown to be ribosome-associated, with an inhibitory effect on translation. tRF(Val) resulting from tRNA-Val cuts are the most abundant tRNA fragments interacting with the small subunit of the ribosome, resulting in the inhibition of the translation stage through interference with peptidyl transferase activity. It was also observed that under conditions of stress, tRF(Val) is strongly induced, suggesting a regulatory role for tRF in archaea, as had been shown in eukaryotes (Gebetsberger et al. 2012). Ultimately, tRF formation appears to be an equally widespread phenomenon in archaea, with a fairly wide diversity of cut sites along the tRNA sequence, and a preponderance of cut sites in the anticodon loop of tRNAs.

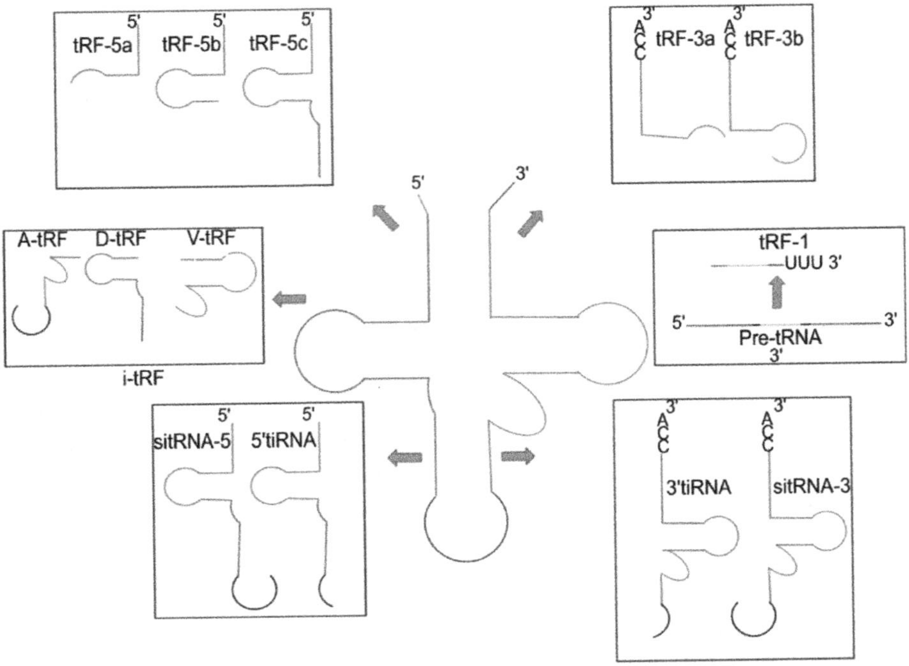

Figure 6.5. *Classification of tRNA-derived small RNAs (according to Xie et al. 2020). For a color version of this figure, see www.iste.co.uk/clouet/archaea2.zip*

COMMENT ON FIGURE 6.5.– *tiRNAs, which include 5'iRNAs, sitRNA-5s, 3'tiRNAs and sitRNA-3s, are derived from mature tRNAs by cleavage in the anticodon loop. sitRNAs are differentiated by length and have been observed under stress conditions. tRF-5 (3 subtypes) and tRF-3 (2 subtypes) are generated from mature 5' and 3' ends, respectively, and are differentiated by size. The i-tRFs, on the other hand, are derived from internal cuts. tRF-1, also known as 3'U-tRF, results from pre-tRNA maturation.*

Since 2005, it has been shown in archaea that certain tRNA genes essential for survival are absent from the genome, at least in the typical form. These tRNA genes are either interrupted by noncanonical intron sequences or separated on two or three genes far apart on the genome (Kanai et al. 2015). This discovery revealed a second pathway for the synthesis of the mature tRNA molecule in archaea. In

Nanoarchaeum equitans, tRNAs, usually encoded by a single gene, are encoded by two distinct, noncontiguous genes on the genome, each carrying the sequence of half a tRNA (split-tRNA). The gene for the 5′ half of the tRNA codes for the first 37 nucleotides of the tRNA, ending directly after the anticodon sequence, in other words, at the usual intron position. The two half-tRNAs are then linked together by a ligation reaction (Randau et al. 2005). The exact reason for these two mechanisms of tRNA biosynthesis is not yet known, but an explanation linked to the molecular evolution of tRNAs has been put forward (Di Giulio 2009; Kanai 2015).

6.2.5. *Noncoding RNA introns in archaea*

The presence of introns in tRNA genes has been demonstrated in all three domains of life, with 15% of archaeal tRNAs possessing an intron (Sugahara et al. 2008). Although tRNA introns are quite short – between 11 and 129 nucleotides – their presence in pre-tRNA disrupts the typical structural organization of mature tRNAs. Intron splicing of pre-tRNAs is therefore essential and, in archaea, involves a different mechanism to that described for eukaryotic mRNA splicing by the "spliceosome". As archaea do not possess a group I self-splicing intron, the pre-tRNA is recognized by a specific splicing complex including an endonuclease EndA, which will recognize a particular pseudosymmetric RNA structure present at the junction of the intron/exon sequences. This structural motif, known as BHB, is composed of a helix with bulges and is the motif recognized by the splicing endonuclease (Fujishima et al. 2011). After cleavage by the endonuclease EndA, the exons encoding the two half-tRNAs are ligated by an RNA ligase of the RtcB family, while the intron is released, after circularization, by another RNA ligase family (Becker et al. 2017; Clouet-d'Orval et al. 2018). The majority of pre-tRNAs only contain one intron adjacent to the anticodon sequence. In other words, between nucleotides 37 and 38 of the tRNA (canonical position), but some introns may be present several times and at unconventional positions, such as the D, TΨ or variable loops of pre-tRNAs (Marck 2003). The intron present in the pre-tRNA Trp anticodon is the most extensively studied and best characterized in several archaeal species (*P. abyssi* and *H. volcanii*). This intron features two particularities: (i) the presence of box C/D RNA motifs in its sequence, capable of guiding the methylation reaction on the ribose nucleotides of the anticodon loop; (ii) a circular RNA (circRNA) structure with no free ends (see sections on box C/D RNAs and circRNAs for further details).

Ribosome biogenesis involves multiple stages of rRNA maturation and protein assembly. Genes encoding rRNAs (16S, 23S and 5S rDNAs) are generally associated in operons which, after transcription, will form precursors (pre-rRNAs) to be matured, leading to functional rRNAs. Several introns, ranging from 55 to nearly 600 nucleotides, have been identified in conserved regions of the 16S and 23S pre-rRNAs of several archaea in the order *Desulfurococcales* and *Thermoproteales* (Lykke-Andersen 1997; Tocchini-Valentini et al. 2011). As with the pre-tRNAs described above, it is the "BHB" motif, present at the junction of the intron/exon sequences, that will be the cutting site for the specific endonuclease (EndA), but other exonucleases and endonucleases are also involved in the maturation process. The rRNA fragments generated after hydrolysis will then undergo ligation steps to transiently form pre-16S and pre-23S circRNAs, before being hydrolyzed and undergoing final maturation to generate functional rRNAs (Tang 2002). Certain introns, however, have been localized at functional ribosome centers and could encode hypothetical proteins, after excision and potential circularization (Dalgaard and Garrett 1992). Further characterization of introns in other archaeal species is required to confirm these results or to classify these functions as isolated cases.

6.3. ncRNA identification

6.3.1. *Experimental approaches*

The identification of transcriptomes is now possible thanks to new-generation sequencing techniques. Since 2004, it has been possible to obtain the human genome in an automated, miniaturized and rapid way, and at a much lower cost than with previous Sanger-type techniques. There are three improvements incorporated in next-generation sequencing: (i) preparation of DNA libraries without the need for cloning DNA fragments in bacterial vectors; (ii) simultaneous performance of hundreds to thousands of sequencing reactions; (iii) direct reading of sequencing results (by fluorescence, semi-conductivity, etc.) without the need for an electrophoresis step. Today, RNA sequencing (RNA-seq) enables detailed and nearly exhaustive characterization of each transcriptome, despite the great diversity of RNA molecules and differences in abundance between RNA families (Figure 6.6(a)). RNA libraries for sequencing are prepared while considering the high abundance of rRNAs, the differential expression level between RNAs and the multiple specificities of ncRNAs (high diversity, highly variable size, low abundance for some of them). RNA fractionation or capture of RNA subcategories is therefore performed in order to obtain the most exhaustive possible representation of the RNA sample to be analyzed. Typically, RNA isolated from a sample undergoes gentle cutting if its size is too large, before being fused to adapters containing the sequences required for immobilization of the DNA fragments on a solid surface, as well as for the sequencing step itself. After reverse-transcription and PCR amplification of the cDNA, sequencing is carried out using a range of current techniques (Illumina, 454, Ion Torrent, SOLiD, etc.) (Figure 6.6(b)). Following a base-calling step, aimed at specifically associating the sequencing signal with a base, the sequenced fragments are obtained in the form of base sequences, known as reads, which are then subjected to bioinformatics analysis.

a) b)

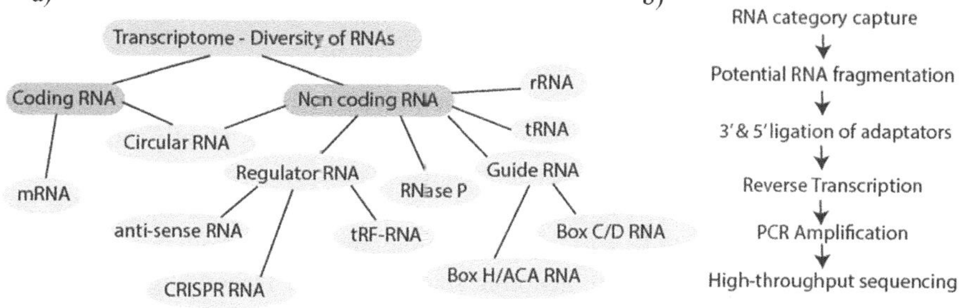

c)

Subcategories of RNA	Approach	Type of RNA-seq
Total RNA	Higher functional coverage after rRNA depletion	Ribo-zero
mRNA	Polysomes associated mRNA - Ultracentrifugation mRNA protected from RNases by the ribosome	Ribosome-Profiling Ribo-seq
Small RNA	Size selection - Electrophoresis gel	Small RNA-seq
Protein-assoociated RNA	RNA immunoprecipitation Covalent RNA-protein cross-link	RIP-Seq CLIP-Seq
Methylated RNA	Immunoprecipitation of methylated RNA (e.g.: m6A)	Me-RIP-Seq
Circular RNA	Resistance to RNase R treatment	circRNA-Seq

Figure 6.6. *Development of sequencing techniques adapted to each RNA subcategory. (a) Representation of the diversity of the RNA world; (b) procedure to follow, from RNA preparation from a sample to high-throughput sequencing, including preparation of sequencing libraries; (c) sequencing techniques adapted to a specific RNA subcategory. The approach for isolating an RNA population is indicated*

New methods of sample preparation for sequencing have been developed to specifically access sub-populations of RNA involved in specific biological processes, commonly referred to as "RNA category capture". Figure 6.6(c) summarizes the main RNA-seq techniques used to identify RNA sub-populations, including fractionation of total RNA (Ribo-zero, small RNA-seq), immunoprecipitation of RNA specifically bound to a target protein (RIP-seq), immunoprecipitation of RNA carrying a methyl-type modified nucleotide (Me-RIP-seq), chemical or photoactivation bridging of an RNA-protein complex

(CLIP-seq), selection of mRNAs in "active translation" and specifically bound to ribosomes (Ribo-seq, Ribosome-profiling), or enrichment of circRNAs thanks to their resistance to exoribonucleases (circRNA-seq).

6.3.2. *Bioinformatics approaches*

The democratization of genome and transcriptome sequencing under a variety of conditions, and the recognition of the major role played by ncRNAs in a wide range of processes, have generated growing interest in the identification and characterization of associated genes. In archaea, as in eukaryotes and bacteria, unlike protein-coding RNAs, the absence of a sequence signature shared by all ncRNAs makes it impossible to identify and annotate them exhaustively using a single computerized method, at the genomic sequence level. Depending on whether the aim is to identify ncRNAs that are members of known families or those belonging to new families to be characterized, approaches differ (Figure 6.7). On a genome-wide scale, the in silico identification and annotation of ncRNAs belonging to known families has led to the development of numerous methods, exploiting a diversity of signals that are specific to each ncRNA family characterized. As early as the 1990s, the conservation of the secondary structure of a given RNA family was exploited to develop the first approaches for identifying the vast majority of members of well-characterized families. Quite naturally, the first dedicated tools targeted tRNAs (Fichant and Burks 1991; Pavesi et al. 1994). The tRNAscan-SE tool (Lowe and Eddy 1997) is now the standard for genome annotation. It was only later that sequence conservation, specific to rRNAs, was exploited to give rise to the RNAmmer tool (Lagesen et al. 2007). Ultimately, the discovery and characterization of box C/D and H/ACA snoRNAs in eukaryotes contributed to their identification in archaea and to the development of new specific or more general approaches, targeting the characteristic motifs of these genes (Omer 2000; Thébault et al. 2006), but also their complementary sequence with a target RNA. On a broader scale, through available covariance models that integrate the structural conservation of members of a given family, the RFAM database (Kalvari et al. 2018a, 2018b) and the Infernal suite (Nawrocki and Eddy 2013) are the first developments to enable structural and functional annotation of a wide range of ncRNAs at the whole-genome scale.

The discovery of new ncRNA families has long been limited to the identification of conserved intergenic regions between closely related organisms or, more anecdotally, to the exploitation of a compositional bias in certain organisms (Klein et al. 2002; Schattner 2002). The identification of conserved intergenic regions has been successfully applied to genomes that are close enough to identify similarities, but have also sufficiently diverged to detect covariations in potentially structured regions. This type of approach, initially developed to detect ncRNAs in the yeast *S. cereviae* (Olivas 1997) or in bacterial genomes (Argaman et al. 2001), has given rise to generic approaches (Rivas and Eddy 2001) that were used and adapted in archaea (Phok et al. 2011).

More recently, in several archaea species, the discovery of new families has been considerably accelerated by the bioinformatic analysis of transcripts derived from the production of transcriptomes of various sizes and conditions. Like in eukaryotes, these large-scale approaches provided access to an unexpected diversity and quantity of ncRNAs including, for example, circRNAs, asRNAs, sRNAs, lncRNAs and tsRNAs (Jager et al. 2009; Wurtzel et al. 2010; Bernick et al. 2012b; Danan et al. 2012; Toffano-Nioche et al. 2013; Babski et al. 2016; Becker et al. 2017; Gelsinger and DiRuggiero 2018b; Orell et al. 2018). The data resulting from the sequencing of a transcriptome are made up of a very large number (potentially several tens of millions) of reads that may contain errors, and whose position and orientation on their reference genome remain to be determined by global and nearly exact alignment of reads on the reference genome.

New methods, which are far more efficient for this type of alignment than those implemented in traditional alignment tools such as Blastn (Altschul et al. 1990), have been developed and are now available in the form of tools such as, among the best known, Bowtie (Langmead et al. 2009) and BWA (Li and Durbin 2009). For regions covered by these transcripts, but which have not yet been described, the search for ORFs can then be used to determine their coding or noncoding potential. However, given the specific natures of many of these ncRNAs (circularization, split-RNA, etc.), the availability of tools for the exhaustive identification of ncRNAs from archaeal transcriptomes remains a challenge.

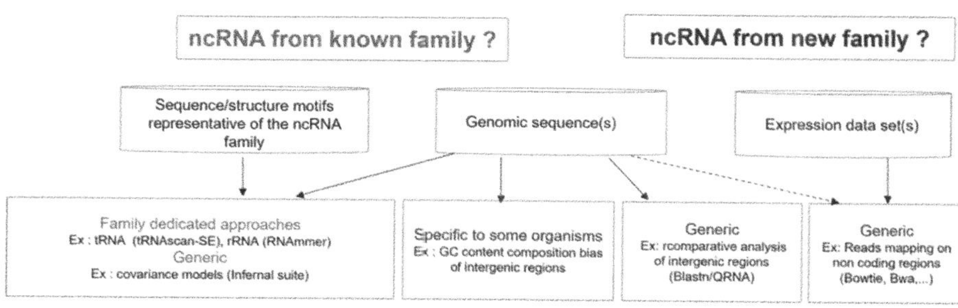

Figure 6.7. *Classification of bioinformatics approaches for identifying noncoding RNA in archaea. For a color version of this figure, see www.iste.co.uk/clouet/archaea2.zip*

COMMENT ON FIGURE 6.7.– *When the genome sequence is available, members of known archaeal families can be identified using conventional ncRNA identification and annotation approaches (green). To identify new ncRNA families (in blue), it is necessary to have access to several nearby genomes to apply comparative approaches, or to one or more transcriptomes of the target organism to access transcripts from regions that have not yet been annotated. Certain classes of ncRNAs, such as antisense RNAs, can only be identified by sequencing one or more transcriptomes.*

6.4. Circularization of noncoding RNAs

6.4.1. *Overview of circRNAs*

A new category of RNA, known as circRNA, has been identified in recent years, mainly thanks to next-generation sequencing techniques on transcriptomes that are non-polyadenylated or circRNA enriched. circRNAs are RNA molecules lacking free 5′-phosphate and 3′-OH ends, giving them a circular structure that is resistant to exoribonucleases compared to their linear counterparts. Today, circRNAs are described as ubiquitous and present in all three domains of life (bacteria, archaea and eukaryotes), whereas just ten years ago, they were perceived as being of low abundance and as aberrant products of pre-RNA maturation processes. Early sequencing studies only enabled partial identification of circRNAs, as these techniques used a poly(A)-tail RNA selection step during preparation of the RNA library, which automatically excluded circRNAs lacking a poly(A) tail. In addition, bioinformatics analyses of RNA-seq results were based on linear alignments of sequencing reads on the genome which, again, was unfavorable for the identification of circRNAs for which the alignment of reads is reversed on the genome, particularly at junction sequences, known as "circularization" (Becker et al. 2017).

circRNAs have been identified in several organisms and for several RNA families, such as (i) the circular single-stranded RNA genome of plant viroids; (ii) excised circular introns of pre-tRNAs or group II introns; (iii) circular rRNA maturation or transfer intermediates; (iv) various noncoding circRNAs involved in the regulation of transcription, spiking and translation steps (microRNA sponges) (Lasda and Parker 2014; Chen 2016; Wilusz 2018). The high number and diversity of circRNAs identified in humans has initiated the scientific community's interest in circRNAs, with several thousand circRNAs observed, with variable expression levels specific to certain tissues, age or pathologies. Today, circRNAs are being considered as new potential biomarkers for the diagnosis and treatment of pathologies (Qu et al. 2017).

In archaea, the first circRNAs were identified a few years before the era of next-generation sequencing. These were circRNAs belonging to several functional categories, such as the excised intron of tRNA-Trp from the halophilic archaea *Haloferax volcanii* (Clouet-d'Orval et al. 2001; Salgia 2003). This circular intron of tRNA-Trp has the particularity of containing a box C/D RNA sequence that can guide the methylation of nucleotides in the anticodon loop of mature tRNA-Trp (Clouet-d'Orval et al. 2001). Other circular box C/D RNAs have been isolated from *P. furiosus*, within the native ribonucleoprotein complex involved in rRNA methylation, suggesting that circular box C/D RNAs may act as guides for nucleotide modifications in *P. furiosus* (Starostina et al. 2004). In addition, several circRNAs derived from 16S and 23s rRNAs of hyperthermophilic archaea have been described as circular rRNA maturation intermediates (Dalgaard and Garrett 1992; Lykke-Andersen and Garrett 1994; Tang 2002).

The list of circRNAs has been considerably extended using high-throughput sequencing techniques and shows coherence on the classification of circRNAs in several archaea (*P. abyssi*, *P. furiosus*, *S. solfataricus* and *N. equitans*). Five types of RNA are frequently identified as circular in these transcriptomes: box C/D ncRNAs, tRNA introns, SRP RNA, RNaseP RNA and rRNA precursors. However, it is still currently difficult to comprehensively identify all circRNAs, suggesting that the number of circRNAs is still underestimated in archaea (Becker et al. 2019).

6.4.2. *Biogenesis of circRNAs*

The synthesis of several circRNAs has been described in archaea as a spliceosome-independent pathway, involving an RNA ligase enzyme of the Rnl3 family. This mechanism differs from that described in eukaryotes, where a reverse splicing mechanism known as backsplicing is observed. This mechanism, handled by the spliceosome in eukaryotes, is due to an inverted arrangement of splice sites compared with the usual arrangement for canonical splicing.

In the case of the reverse splicing reaction, the 5′-donor splice site is located downstream of the 3′-acceptor splice site, which generates a circRNA (Figure 6.8(a)) (Ashwal-Fluss et al. 2014). Linear and reverse splicing mechanisms can compete for the same RNA in a cell, with an increase in backsplicing rate when the circularized region is flanked by sequences favoring 3D juxtaposition of splice sites by base-pairing. These linking elements can be complementary repeat sequences, protein-binding motifs, or derived from exon skipping mechanisms (Wilusz 2018).

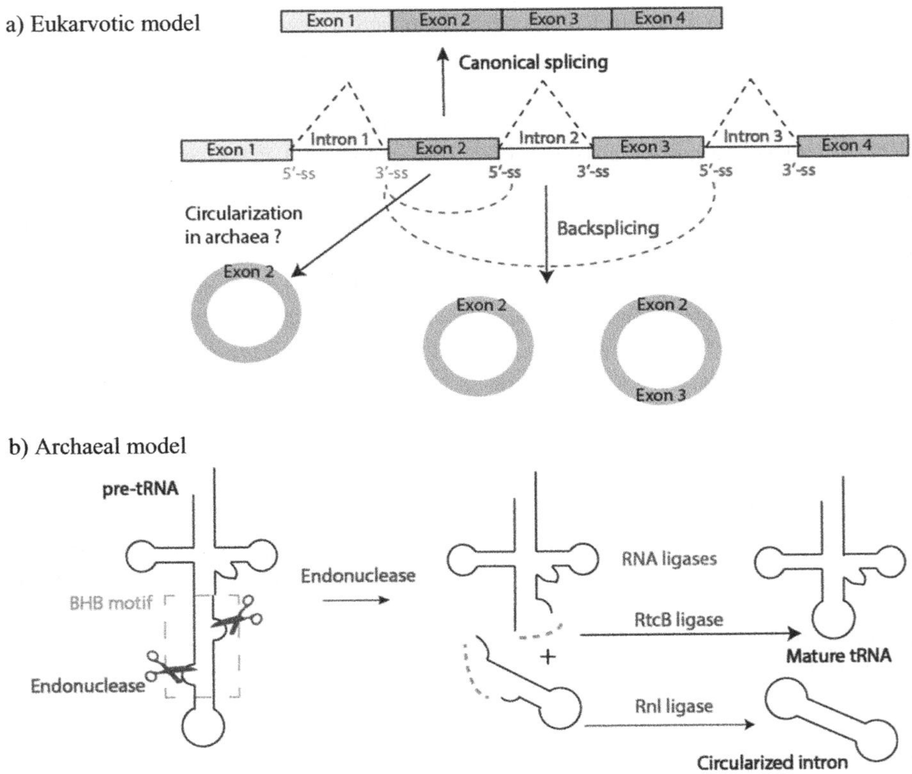

Figure 6.8. *Circular RNA biogenesis. For a color version of this figure, see www.iste.co.uk/clouet/archaea2.zip*

COMMENT ON FIGURE 6.8.– *(a) In eukaryotes, backsplicing is used, shown in red dotted lines. Circular RNAs can be derived from both exonic and intronic sequences. The canonical splicing pathway is indicated by black dotted lines; (b) example of intron circularization of a tRNA precursor in archaea. After excision of the intron, it is circularized by an RNA ligase other than RtcB ligase, and the two tRNA halves are ligated to form the mature tRNA.*

The mechanism of RNA circularization in archaea is enzyme-dependent and not related to a backsplicing mechanism, since splicing is limited to a few RNAs in archaea. Nevertheless, it has been described that 15% of archaeal tRNA genes seem to contain one or more introns, with varying sizes and insertion areas in the tRNA gene (Marck 2003). Introns are eliminated through the action of a splicing endonuclease that recognizes a structural motif that is common to all subsets, the

BHB motif (Tocchini-Valentini et al. 2005a). Splicing endonucleases exist in different architectures, but will always cut, at the exon-intron junction, between the two bulges (Tocchini-Valentini et al. 2005b). The excised intron will then be circularized, in parallel with the ligation of the two tRNA halves, to form the mature tRNA (Clouet-d'Orval et al. 2001; Salgia 2003) (Figure 6.8(b)). Similarly, the 16S and 23S pre-rRNAs will be circularized close to the "BHB" motifs after endonuclease cleavage. The first RNA ligases involved in the RNA circularization reaction have been characterized, belonging to the new Rnl3 family of RNA ligases described in the archae *Pyrococcus abyssi* (PDB ID 2VUG) (Brooks et al. 2008) and *Methano-bacterium thermoautotrophicum* (PDB ID 5D1P) (Torchia et al. 2008).

Unlike other RNA ligases, which are monomeric, Rnl3 RNA ligases have an atypical structure in heterodimeric form, with a characteristic dimerization domain (Becker et al. 2019). Each subunit contains a nucleotidyl-transferase-like catalytic domain, with no current indication whether the active sites are independent or whether they jointly bind and circularize the same RNA molecule. The phylogenetic distribution of RNA ligases Rnl3 show that they are present in many archaea, with the identification of 97 homologous sequences, mainly in thermophiles and hyperthermophiles, that have an optimal growth temperature between 50°C and 95°C (Becker et al. 2019). In the archaea *P. abyssi*, it was recently shown that the RNA ligase Pab 1020, from the Rnl3 family, was capable of circularizing physiological RNAs in vitro, such as box C/D RNAs or 5S rRNA (Becker et al. 2017).

6.5. Conclusion

In recent years, a large number of ncRNAs have been discovered in archaea of various phyla (Euryarchaea, Crenarchaea and Nanoarchaea), mainly thanks to new high-throughput sequencing techniques. For many of them, their function has been clarified in relation to the biological processes of transcriptional and translational regulation. To date, no class of ncRNA has been identified that is specific to archaea. All of the RNA classes identified are found in bacteria or eukaryotes.

Characterization of ncRNAs in archaea remains limited when compared to eukaryotes and bacteria. The limiting factor is that very few archaea species can be cultivated in the laboratory. As with eukaryotes and bacteria, the diversity of archaeal transcriptome data and, more generally, that of "omics" data will, in time, enable us to enrich the repertoire of ncRNAs and characterize them, with the aim of better understanding their roles in extreme environments.

6.6. References

Altschul, S.F., Gish, W., Miller, W., Myers, E.W., Lipman, D.J. (1990). Basic local alignment search tool. *J. Mol. Biol.*, 215(3), 403–410.

Argaman, L., Hershberg, R., Vogel, J., Bejerano, G., Wagner, E.G.H., Margalit, H., Altuvia, S. (2001). Novel small RNA-encoding genes in the intergenic regions of *Escherichia coli*. *Curr. Biol.*, 11(12), 941–950.

Ashwal-Fluss, R., Meyer, M., Pamudurti, N.R., Ivanov, A., Bartok, O., Hanan, M., Evantal, N., Memczak, S., Rajewsky, N., Kadener, S. (2014). circRNA biogenesis competes with pre-mRNA splicing. *Mol. Cell.*, 56(1), 55–66.

Babski, J., Maier, L.-K., Heyer, R., Jaschinski, K., Prasse, D., Jäger, D., Randau, L., Schmitz, R.A., Marchfelder, A., Soppa, J. (2014). Small regulatory RNAs in Archaea. *RNA Biol.*, 11(5), 484–493.

Babski, J., Haas, K.A., Näther-Schindler, D., Pfeiffer, F., Förstner, K.U., Hammelmann, M., Hilker, R., Becker, A., Sharma, C.M., Marchfelder, A. et al. (2016). Genome-wide identification of transcriptional start sites in the haloarchaeon *Haloferax volcanii* based on differential RNA-Seq (dRNA-Seq). *BMC Genomics*, 17, 629.

Bachellerie, J.P., Cavaillé, J., Hüttenhofer, A. (2002). The expanding snoRNA world. *Biochimie*, 84(8), 775–790.

Baker, D.L. (2005). RNA-guided RNA modification: Functional organization of the archaeal H/ACA RNP. *Genes. Dev.*, 19(10), 1238–1248.

Becker, H.F., Héliou, A., Djaout, K., Lestini, R., Regnier, M., Myllykallio, H. (2017). High-throughput sequencing reveals circular substrates for an archaeal RNA ligase. *RNA Biol.*, 14(8), 1075–1085.

Becker, H.F., L'Hermitte-Stead, C., Myllykallio, H. (2019). Diversity of circular RNAs and RNA ligases in archaeal cells. *Biochimie*, 164, 37–44.

Bernick, D.L., Dennis, P.P., Lui, L.M., Lowe, T.M. (2012a). Diversity of antisense and other non-coding RNAs in archaea revealed by comparative small RNA sequencing in four Pyrobaculum species. *Front. Microbiol.*, 3, 231.

Bernick, D.L., Dennis, P.P., Hochsmann, M., Lowe, T.M. (2012b). Discovery of Pyrobaculum small RNA families with atypical pseudouridine guide RNA features. *RNA*, 18(3), 402–411.

Brooks, M.A., Meslet-Cladiére, L., Graille, M., Kuhn, J., Blondeau, K., Myllykallio, H., van Tilbeurgh, H. (2008). The structure of an archaeal homodimeric ligase which has RNA circularization activity. *Protein Sci.*, 17(8), 1336–1345.

Charron, C., Manival, X., Cléry, A., Senty-Ségault, V., Charpentier, B., Marmier-Gourrier, N., Branlant, C., Aubry, A. (2004). The archaeal sRNA binding protein L7Ae has a 3D structure very similar to that of its eukaryal counterpart while having a broader RNA-binding specificity. *J. Mol. Biol.*, 342(3), 757–773.

Chen, L.-L. (2016). The biogenesis and emerging roles of circular RNAs. *Nat. Rev. Mol. Cell Biol.*, 17(4), 205–211.

Clouet-d'Orval, B., Bortolin, M.L., Gaspin, C., Bachellerie, J.P. (2001). Box C/D RNA guides for the ribose methylation of archaeal tRNAs. The tRNATrp intron guides the formation of two ribose-methylated nucleosides in the mature tRNATrp. *Nucleic Acids Res.*, 29(22), 4518–4529.

Clouet-d'Orval, B., Batista, M., Bouvier, M., Quentin, Y., Fichant, G., Marchfelder, A., Maier, L.-K. (2018). Insights into RNA-processing pathways and associated RNA-degrading enzymes in Archaea. *FEMS Microbiol. Rev.*, 42(5), 579–613.

Dalgaard, J.Z. and Garrett, R.A. (1992). Protein-coding introns from the 23S rRNA-encoding gene form stable circles in the hyperthermophilic archaeon *Pyrobaculum organotrophum*. *Gene*, 121(1), 103–110.

Danan, M., Schwartz, S., Edelheit, S., Sorek, R. (2012). Transcriptome-wide discovery of circular RNAs in Archaea. *Nucleic Acids Res.*, 40(7), 3131–3142.

Dennis, P.P., Omer, A., Lowe, T. (2001). A guided tour: Small RNA function in Archaea: Small RNA function in Archaea. *Mol. Microbiol.*, 40(3), 509–519.

Di Giulio, M. (2009). Formal proof that the split genes of tRNAs of *Nanoarchaeum equitans* are an ancestral character. *J. Mol. Evol.*, 69(5), 505–511.

Fichant, G.A. and Burks, C. (1991). Identifying potential tRNA genes in genomic DNA sequences. *J. Mol. Biol.*, 220(3), 659–671.

Fujishima, K., Sugahara, J., Miller, C.S., Baker, B.J., Di Giulio, M., Takesue, K., Sato, A., Tomita, M., Banfield, J.F., Kanai, A. (2011). A novel three-unit tRNA splicing endonuclease found in ultrasmall Archaea possesses broad substrate specificity. *Nucleic Acids Res.*, 39(22), 9695–9704.

Gaspin, C., Cavaillé, J., Erauso, G., Bachellerie, J.-P. (2000). Archaeal homologs of eukaryotic methylation guide small nucleolar RNAs: Lessons from the Pyrococcus genomes 1. *J. Mol. Biol.*, 297(4), 895–906.

Gebetsberger, J. and Polacek, N. (2013). Slicing tRNAs to boost functional ncRNA diversity. *RNA Biol.*, 10(12), 1798–1806.

Gebetsberger, J., Zywicki, M., Künzi, A., Polacek, N. (2012). tRNA-derived fragments target the ribosome and function as regulatory non-coding RNA in *Haloferax volcanii*. *Archaea*, 2012, 1–11.

Gebetsberger, J., Wyss, L., Mleczko, A.M., Reuther, J., Polacek, N. (2017). A tRNA-derived fragment competes with mRNA for ribosome binding and regulates translation during stress. *RNA Biol.*, 14(10), 1364–1373.

Gelsinger, D. and DiRuggiero, J. (2018a). The non-coding regulatory RNA revolution in Archaea. *Genes*, 9(3), 141.

Gelsinger, D.R. and DiRuggiero, J. (2018b). Transcriptional landscape and regulatory roles of small noncoding RNAs in the oxidative stress response of the haloarchaeon *Haloferax volcanii*. *J. Bacteriol.*, 200(9), e00779-17.

Gomes-Filho, J.V., Daume, M., Randau, L. (2018). Unique archaeal small RNAs. *Annu. Rev. Genet.*, 52, 465–487.

Hamma, T., Reichow, S.L., Varani, G., Ferré-D'Amaré, A.R. (2005). The Cbf5–Nop10 complex is a molecular bracket that organizes box H/ACA RNPs. *Nat. Struct. Mol. Biol.*, 12, 1101–1107.

Heyer, R., Dörr, M., Jellen-Ritter, A., Späth, B., Babski, J., Jaschinski, K., Soppa, J., Marchfelder, A. (2012). High throughput sequencing reveals a plethora of small RNAs including tRNA derived fragments in *Haloferax volcanii*. *RNA Biol.*, 9(7), 1011–1018.

Hüttenhofer, A. and Schattner, P. (2006). The principles of guiding by RNA: Chimeric RNA–protein enzymes. *Nat. Rev. Genet.*, 7, 475–482.

Jäger, D., Sharma, C.M., Thomsen, J., Ehlers, C., Vogel, J., Schmitz, R.A. (2009). Deep sequencing analysis of the *Methanosarcina mazei* Gö1 transcriptome in response to nitrogen availability. *Proc. Natl. Acad. Sci.*, 106(51), 21878–21882.

Jäger, D., Pernitzsch, S.R., Richter, A.S., Backofen, R., Sharma, C.M., Schmitz, R.A. (2012). An archaeal sRNA targeting cis- and trans-encoded mRNAs via two distinct domains. *Nucleic Acids Res.*, 40(21), 10964–10979.

Kalvari, I., Nawrocki, E.P., Argasinska, J., Quinones-Olvera, N., Finn, R.D., Bateman, A., Petrov, A.I. (2018a). Non-coding RNA analysis using the Rfam database. *Curr. Protoc. Bioinforma.*, 62(1), e51.

Kalvari, I., Argasinska, J., Quinones-Olvera, N., Nawrocki, E.P., Rivas, E., Eddy, S.R., Bateman, A., Finn, R.D., Petrov, A.I. (2018b). Rfam 13.0: Shifting to a genome-centric resource for non-coding RNA families. *Nucleic Acids Res.*, 46(D1), D335–D342.

Kanai, A. (2015). Disrupted tRNA genes and tRNA fragments: A perspective on tRNA gene evolution. *Life*, 5(1), 321–331.

Klein, R.J., Misulovin, Z., Eddy, S.R. (2002). Noncoding RNA genes identified in AT-rich hyperthermophiles. *Proc. Natl. Acad. Sci.*, 99(11), 7542–7547.

Lagesen, K., Hallin, P., Rødland, E.A., Stærfeldt, H.-H., Rognes, T., Ussery, D.W. (2007). RNAmmer: Consistent and rapid annotation of ribosomal RNA genes. *Nucleic Acids Res.*, 35(9), 3100–3108.

Langmead, B., Trapnell, C., Pop, M., Salzberg, S.L. (2009). Ultrafast and memory-efficient alignment of short DNA sequences to the human genome. *Genome Biol.*, 10, R25.

Lasda, E. and Parker, R. (2014). Circular RNAs: Diversity of form and function. *RNA*, 20(12), 1829–1842.

Li, H. and Durbin, R. (2009). Fast and accurate short read alignment with Burrows–Wheeler transform. *Bioinformatics*, 25(14), 1754–1760.

Lowe, T.M. and Eddy, S.R. (1997). tRNAscan-SE: A program for improved detection of transfer RNA genes in genomic sequence. *Nucleic Acids Res.*, 25(5), 955–964.

Lykke-Andersen, J. (1997). Archaeal introns: Splicing, intercellular mobility and evolution. *Trends Biochem. Sci.*, 22(9), 326–331.

Lykke-Andersen, J. and Garrett, R.A. (1994). Structural characteristics of the stable RNA introns of archaeal hyperthermophiles and their splicing junctions. *J. Mol. Biol.*, 243(5), 846–855.

Makarova, K.S., Wolf, Y.I., Alkhnbashi, O.S., Costa, F., Shah, S.A., Saunders, S.J., Barrangou, R., Brouns, S.J.J., Charpentier, E., Haft, D.H. et al. (2015). An updated evolutionary classification of CRISPR–Cas systems. *Nat. Rev. Microbiol.*, 13, 722–736.

Marchfelder, A., Fischer, S., Brendel, J., Stoll, B., Maier, L.-K., Jäger, D., Prasse, D., Plagens, A., Schmitz, R.A., Randau, L. (2012). Small RNAs for defence and regulation in archaea. *Extremophiles*, 16(5), 685–696.

Marck, C. (2003). Identification of BHB splicing motifs in intron-containing tRNAs from 18 archaea: Evolutionary implications. *RNA*, 9(12), 1516–1531.

Nawrocki, E.P. and Eddy, S.R. (2013). Infernal 1.1: 100-Fold faster RNA homology searches. *Bioinformatics*, 29(22), 2933–2935.

Nolivos, S. (2005). The K-loop, a general feature of the Pyrococcus C/D guide RNAs, is an RNA structural motif related to the K-turn. *Nucleic Acids Res.*, 33(20), 6507–6514.

Olivas, W. (1997). Analysis of the yeast genome: Identification of new non-coding and small ORF-containing RNAs. *Nucleic Acids Res.*, 25(22), 4619–4625.

Omer, A.D. (2000). Homologs of small nucleolar RNAs in Archaea. *Science*, 288(5465), 517–522.

Omer, A.D., Ziesche, S., Decatur, W.A., Fournier, M.J., Dennis, P.P. (2003). RNA-modifying machines in archaea: Homologues of snoRNPs in Archea. *Mol. Microbiol.*, 48(3), 617–629.

Orell, A., Tripp, V., Aliaga-Tobar, V., Albers, S.-V., Maracaja-Coutinho, V., Randau, L. (2018). A regulatory RNA is involved in RNA duplex formation and biofilm regulation in *Sulfolobus acidocaldarius*. *Nucleic Acids Res.*, 46(9), 4794–4806.

Pavesi, A., Conterio, F., Bolchi, A., Dieci, G., Ottonello, S. (1994). Identification of new eukaryotic tRNA genes in genomic DNA databases by a multistep weight matrix analysis of transcriptional control regions. *Nucleic Acids Res.*, 22(7), 1247–1256.

Phok, K., Moisan, A., Rinaldi, D., Brucato, N., Carpousis, A.J., Gaspin, C., Clouet-d'Orval, B. (2011). Identification of CRISPR and riboswitch related RNAs among novel noncoding RNAs of the euryarchaeon *Pyrococcus abyssi*. *BMC Genomics*, 12, 312.

Qu, S., Zhong, Y., Shang, R., Zhang, X., Song, W., Kjems, J., Li, H. (2017). The emerging landscape of circular RNA in life processes. *RNA Biol.*, 14(8), 992–999.

Randau, L., Münch, R., Hohn, M.J., Jahn, D., Söll, D. (2005). *Nanoarchaeum equitans* creates functional tRNAs from separate genes for their 5′- and 3′-halves. *Nature*, 433, 537–541.

Reichow, S.L., Hamma, T., Ferre-D'Amare, A.R., Varani, G. (2007). The structure and function of small nucleolar ribonucleoproteins. *Nucleic Acids Res.*, 35(5), 1452–1464.

Renalier, M.-H., Joseph, N., Gaspin, C., Thebault, P., Mougin, A. (2005). The Cm56 tRNA modification in archaea is catalyzed either by a specific 2'-O-methylase, or a C/D sRNP. *RNA*, 11(7), 1051–1063.

Rivas, E. and Eddy, S.R. (2001). Noncoding RNA gene detection using comparative sequence analysis. *BMC Bioinformatics*, 2(8), 1–19.

Rozhdestvensky, T.S. (2003). Binding of L7Ae protein to the K-turn of archaeal snoRNAs: A shared RNA binding motif for C/D and H/ACA box snoRNAs in Archaea. *Nucleic Acids Res.*, 31(3), 869–877.

Salgia, S.R. (2003). Two reactions of *Haloferax volcanii* RNA splicing enzymes: Joining of exons and circularization of introns. *RNA*, 9(3), 319–330.

Schattner, P. (2002). Searching for RNA genes using base-composition statistics. *Nucleic Acids Res.*, 30(9), 2076–2082.

Speed, M.C., Burkhart, B.W., Picking, J.W., Santangelo, T.J. (2018). An archaeal fluoride-responsive riboswitch provides an inducible expression system for hyperthermophiles. *Appl. Environ. Microbiol.*, 84(7), e02306-17.

Starostina, N.G., Marshburn, S., Johnson, L.S., Eddy, S.R., Terns, R.M., Terns, M.P. (2004). Circular box C/D RNAs in *Pyrococcus furiosus*. *Proc. Natl. Acad. Sci.*, 101(39), 14097–14101.

Sugahara, J., Kikuta, K., Fujishima, K., Yachie, N., Tomita, M., Kanai, A. (2008). Comprehensive analysis of archaeal tRNA genes reveals rapid increase of tRNA introns in the order thermoproteales. *Mol. Biol. Evol.*, 25(12), 2709–2716.

Tang, T.H. (2002). RNomics in Archaea reveals a further link between splicing of archaeal introns and rRNA processing. *Nucleic Acids Res.*, 30(4), 921–930.

Thébault, P., de Givry, S., Schiex, T., Gaspin, C. (2006). Searching RNA motifs and their intermolecular contacts with constraint networks. *Bioinforma. Bioinformatics*, 22(17), 2074–2080.

Tocchini-Valentini, G.D., Fruscoloni, P., Tocchini-Valentini, G.P. (2005a). Structure, function, and evolution of the tRNA endonucleases of Archaea: An example of sub-functionalization. *Proc. Natl. Acad. Sci.*, 102(25), 8933–8938.

Tocchini-Valentini, G.D., Fruscoloni, P., Tocchini-Valentini, G.P. (2005b). Coevolution of tRNA intron motifs and tRNA endonuclease architecture in Archaea. *Proc. Natl. Acad. Sci.*, 102(43), 15418–15422.

Tocchini-Valentini, G.D., Fruscoloni, P., Tocchini-Valentini, G.P. (2011). Evolution of introns in the archaeal world. *Proc. Natl. Acad. Sci.*, 108(12), 4782–4787.

Toffano-Nioche, C., Ott, A., Crozat, E., Nguyen, A.N., Zytnicki, M., Leclerc, F., Forterre, P., Bouloc, P., Gautheret, D. (2013). RNA at 92°C: The non-coding transcriptome of the hyperthermophilic archaeon *Pyrococcus abyssi*. *RNA Biol.*, 10(7), 1211–1220.

Torchia, C., Takagi, Y., Ho, C.K. (2008). Archaeal RNA ligase is a homodimeric protein that catalyzes intramolecular ligation of single-stranded RNA and DNA. *Nucleic Acids Res.*, 36(19), 6218–6227.

Watkins, N.J. and Bohnsack, M.T. (2012). The box C/D and H/ACA snoRNPs: Key players in the modification, processing and the dynamic folding of ribosomal RNA: Box C/D and H/ACA snoRNPs. *Wiley Interdiscip. Rev. RNA*, 3(3), 397–414.

Wilusz, J.E. (2018). A 360° view of circular RNAs: From biogenesis to functions. *Wiley Interdiscip. Rev. RNA*, 9(4), e1478.

Wurtzel, O., Sapra, R., Chen, F., Zhu, Y., Simmons, B.A., Sorek, R. (2010). A single-base resolution map of an archaeal transcriptome. *Genome Res.*, 20(1), 133–141.

Xie, Y., Yao, L., Yu, X., Ruan, Y., Li, Z., Guo, J. (2020). Action mechanisms and research methods of tRNA-derived small RNAs. *Signal Transduct. Target. Ther.*, 5(1), 109.

7

Translation in Archaea

Emmanuelle SCHMITT and Yves MECHULAM
BIOC, CNRS, Institut Polytechnique de Paris, Palaiseau, France

Archaea are fascinating, not only for their diversity and ability to colonize extreme biotopes but also for their phylogenetic position between bacteria and eukaryotes. Like bacteria, archaea are devoid of nuclei. They can therefore be grouped together under the term "prokaryotes". Nevertheless, the genetic proximity between archaea and eukaryotes has led some authors to propose that the term "prokaryote" should be banned from the scientific vocabulary, as it brings together organisms as far apart as bacteria and archaea (Pace 2006). It was then argued that, from a functional point of view, prokaryotes differed from eukaryotes in that, in the absence of a nucleus, transcription and translation took place in the same compartment, and the two processes could therefore be coupled together (Martin and Koonin 2006). Transcription-translation coupling was discussed very early on in bacteria and is even the basis of sophisticated regulatory mechanisms. For example, in bacteria, transcription attenuation enables the early termination of operon transcription to be coupled to the efficient translation of a guide peptide that "senses" the abundance of an aminoacyl-tRNA in the cell (Yanofsky 1981). *Escherichia coli* complexes between ribosome and RNA polymerase only recently been studied (Demo et al. 2017; Fan et al. 2017; Kohler et al. 2017). However, the precise functional significance of these complexes remains to be confirmed.

In archaea, electron microscopy experiments using lysed *Thermococcus kodakaraensis* cells revealed that most polysomes were connected to DNA strands, showing that mRNA can begin to be translated before synthesis is complete (French et al. 2007). It is therefore extremely likely that transcription and translation are also

Biology of Archaea 2,
coordinated by Béatrice CLOUJET-D'ORVAL, Bruno FRANZETTI and Philippe OGER.

coupled in these organisms. Even so, because the archaeal RNA polymerase and ribosome are eukaryotic, the molecular mechanisms of transcription-translation coupling in archaea will differ from those in bacteria, at least as far as the direct interactions between the transcription and translation machineries are concerned (McGary and Nudler 2013; Blombach et al. 2019). The proximity of eukaryotes and archaea is indeed supported by the proximity of their ribosomes. Ribosomal RNA sequence characteristics of the small subunit (16S or 18S) led to the definition of the archaeal domain and the tree of life (Woese and Fox 1977; Woese et al. 1983, 1990). Since this seminal work, structural and phylogenetic studies have shown that archaeal ribosomal proteins are either universal or shared with eukaryotes (Lecompte et al. 2002; Yutin et al. 2012; Armache et al. 2013; Ban et al. 2014; Coureux et al. 2016) (Table 7.1).

Domain	Sedimentation rate		rRNA	Ribosomal proteins
Bacteria	70S	30S	16S	21 (15u, 6b)
		50S	23S, 5S	33 (18u, 15b)
Eukaryotes	80S	40S	18S	33 (15u, 18e)
		60S	28S, 5,8S, 5S	46 (18u, 28e)
Archaea	70S	30S	16S	25 (15u, 9e, 1a)
		50S	23S, 5S	39 (19u, 20e)

Table 7.1. *Bacterial, eukaryotic and archaeal ribosomes. Ribosomal protein contents are shown for E. coli, S. cerevisiae and P. abyssi. Numbers of universal (u), eukaryotic (e) and bacterial (b) proteins are given for the above organisms. These numbers may vary slightly depending on the organism. A possible archaea-specific protein (a) has recently been identified in P. abyssi (Coureux et al. 2020)*

In contrast, the organization of archaeal messenger RNAs is more like that of bacteria, with polycistronic genes organized into operons, intron-less messenger RNAs with no cap at the 5′ end, and initiation mechanisms based on mRNA recruitment using *Shine-Dalgarno* (SD) sequences or very short 5′ untranslated regions. Thus, archaea use a eukaryote translation machinery on bacterial messenger RNAs. As with the gene transcription stage, this particular situation offers a unique opportunity to study these mechanisms and deduce insights into how these fundamental stages in the decoding of genetic information have evolved in all three domains of life. Lastly, because archaeal functional systems are generally simpler than their eukaryotic counterparts, biochemical and structural studies in archaea can also provide valuable models for understanding eukaryotic processes (Dmitriev et al. 2011; Borck et al. 2012; Perrochia et al. 2013; Lyu and Whitman 2017). Pioneering biochemical studies have characterized various archaeal ribosomes and demonstrated their ability to function in vitro under extreme conditions (Lake et al. 1984; Londei et al. 1986). These in vitro systems continue to be improved (Lo Gullo et al. 2019).

In addition, archaeal ribosomes have played a crucial role in the race for ribosome structure, thanks to their resistance to specific conditions. In particular, the first crystals of a large subunit diffracting X-rays at high resolution came from *Haloarcula marismortui* (Lake et al. 1984). These crystals eventually led to the resolution of the first large-subunit structure at 2.4 Å resolution (Ban et al. 2000).

7.1. Initiating the translation

Initiating protein synthesis requires identification of the initiation codon, which defines the open reading frame (ORF). In all areas of life, the initiation stage involves a macromolecular complex comprising the small subunit of the ribosome, linked to the mRNA in the presence of a specialized initiator transfer RNA aminoacylated by methionine (ARNt$_i^{Met}$) and initiation factors (IF). Once the initiation codon has been selected at site P of the ribosome, the IFs are released, the large subunit is associated and a complete ribosome that is qualified for elongation is formed (Figure 7.1).

However, even if the general principle is the same in all three domains of life, the molecular mechanisms are very different in bacteria, eukaryotes and archaea. In bacteria, mRNAs are not matured. They often carry a SD sequence, containing a GGAGG consensus sequence complementary to the 3′ end of the 16S rRNA small subunit.

They may also be devoid of a SD sequence, or even have no 5′ untranslated region (5′-UTR) at all. The methionylated initiator tRNA is formylated, and the formyl group is crucial for its selection by the initiation complex (Guillon et al. 1993). Only three IFs are involved: IF1, IF2 and IF3 (for a review, see Mechulam et al. 2011 and Rodnina 2018)

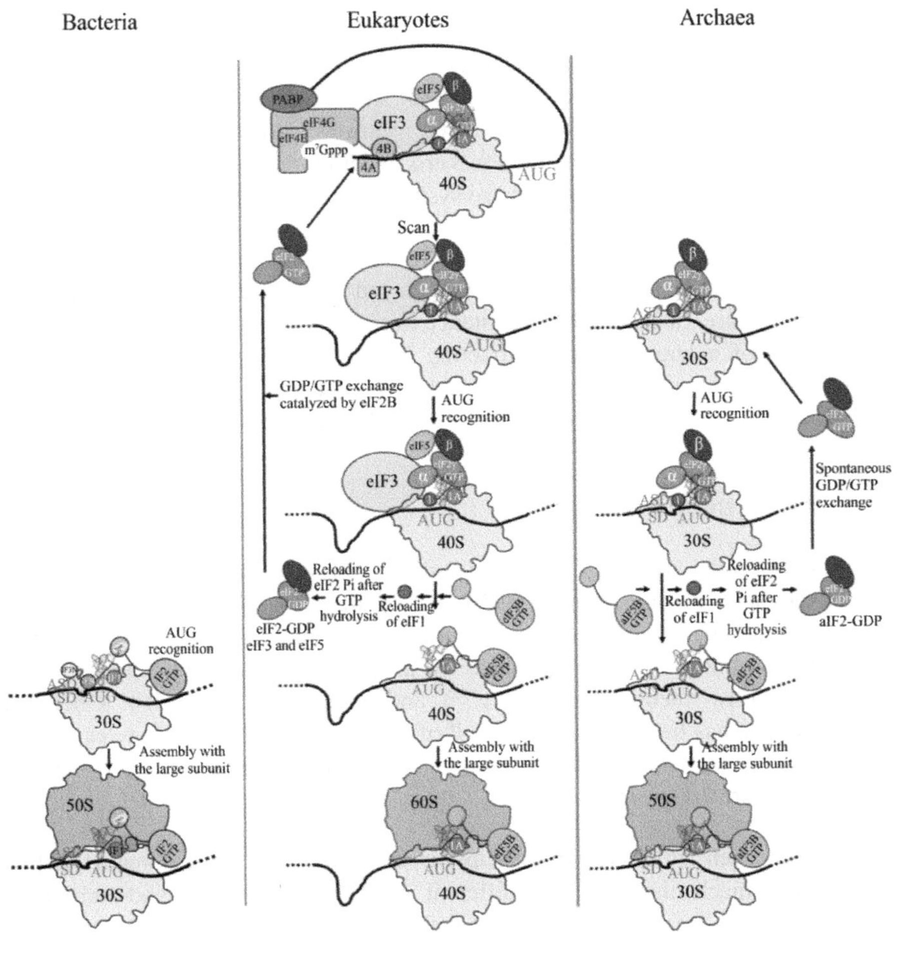

Figure 7.1. *Schematic representation of translation initiation stages in (a) bacteria, (b) eukaryotes and (c) archaea. The e/aIF2 heterotrimer is represented by three colored ovals; blue for the α-subunit, red for the β-subunit and green for the γ-subunit. For simplicity, eIF3 is represented by a single oval, even though this factor contains six subunits in yeast and 13 in mammals. ASD:SD, anti-Shine Dalgarno:Shine Dalgarno duplex. For a color version of this figure, see www.iste.co.uk/clouet/archaea2.zip*

Bacteria	Eukaryotes	Archaea	Function
	mRNA characteristics		
SD Without SD *Leaderless*	Canonical cap-dependent initiation, Kozak motif Non-canonical cap-dependent initiation Non-canonical independent cap initiation IRES-mediated initiation *Leaderless*	SD Without SD *Leaderless*	Recruitment through the initiation complex
	Main characteristics of the initiator tRNA		
Formyl-methionine C_1 A_{72} G_{29}-C_{41},G_{30}-C_{40},G_{31}-C_{39}	Methionine A_1-U_{72} G_{29}-C_{41},G_{30}-C_{40},G_{31}-C_{39}	Methionine A_1-U_{72} G_{29}-C_{41},G_{30}-C_{40},G_{31}-C_{39}	IF2/eIF2 selection Site P
	Initiation factors		
–	eIF2 (α,β,γ)	aIF2 (α,β,γ)	Delivers initiator tRNA to site P
~ IF3*	eIF1	aIF1	Site P
IF1	eIF1A	aIF1A	Site A
IF2	eIF5B	aIF5B	tRNA binding-large subunit joining
–	eIF5	–	GAP
–	eIF2B (α,β,γ,δ,ε)	aIF2B (α,β,δ)#	GEF eIF2B γ,ε
–	eIF3 (6–13 subunits)	–	mRNA binding
–	eIF4F (4A, 4G, 4E)	aIF4A!	5′ cap binding
–	eIF4B	–	Helicase

Table 7.2. *Main characteristics of translation initiation in the three domains of life*

COMMENT ON TABLE 7.2. – *: *IF3 is a two-domain protein. The correspondence between IF3 and e/aIF1 is based on a structural and functional similarity of the C-terminal domain of IF3 with e/aIF1. Despite this similarity, the topologies of the two α–β domains are different. This suggests convergent evolution, rather than* *the presence of a common ancestor. #: The catalytic γ and ε subunits of eIF2B are not found in archaea, and the function of eIF2B homologs in archaea is unclear. !: aIF4A is only present in some archaea. Table based on Schmitt et al. (2019).*

In eukaryotes, the initiation process is more complicated, with many IFs involved (Table 7.1). This stage is also the target of numerous regulations. The mRNAs are matured. They have a cap at the 5′ end and a polyadenylated tail at the 3′ end. The canonical cap-dependent initiation model involves a pre-initiation complex (PIC) containing the small subunit of ribosome, linked to the ternary complex (TC) eIF2-GTP-Met-ARNt$_i^{Met}$, the two small factors eIF1 and eIF1A, and two regulatory proteins eIF5 and eIF3 (Figure 7.1). eIF5 is the GTPase-activating protein of eIF2, and eIF3 is a multimeric factor involved, among other things, in mRNA binding (Table 7.2). In the presence of eIF4 family factors, the PIC is recruited to the capped 5′ end. It then scans the mRNA until an AUG codon in the correct context (Kozak motif) is found. Recognition of the AUG codon stops the scan, triggering the release of factors and the assembly of an 80S complex capable of elongation via the junction with the large subunit, with the help of eIF5B and eIF1A (Figure 7.1) (for a review, see Hinnebusch 2017). Alongside this canonical mechanism, a number of alternative initiation pathways have also been described (Table 7.2) (Shatsky et al. 2018).

In archaea, there is no long-distance scanning, as mRNAs have SD sequences or very short 5′ untranslated regions (Table 7.1). However, genomic analyses have shown that three initiation factors aIF1, aIF1A and aIF2, homologous to the corresponding eukaryotic factors, are present (Figure 7.1 and Table 7.1) (Dennis 1997; Kyrpides and Woese 1998a, 1998b; Benelli and Londei 2011; Gabel et al. 2013). Thus, although there are obvious differences between eukaryotes and archaea, particularly for mRNA recruitment, initiation codon selection is achieved within the same structural core composed of the small ribosomal subunit, mRNA, methionylated initiator tRNA (Met-tRNAt$_i^{Met}$) and the three initiation factors e/aIF1, e/aIF1A and e/aIF2 (Figure 7.2).

7.1.1. *aIF2 and the ternary complex aIF2:GTP:Met-tRNAiMet*

In both archaea and eukaryotes, the methionylated initiator tRNA is transported to the small ribosomal subunit by a GTP-bound heterotrimeric protein (α, β and γ), called aIF2 or eIF2, depending on the living domain from which it originates. This factor specifically binds the methionylated initiator tRNA (Figure 7.2). It recognizes the esterified methionine moiety and prefers an A_1-U_{72} base pair, as found in all eukaryotic or archaeal initiator tRNAs (Kapp and Lorsch 2004; Yatime et al. 2004; Pedulla et al. 2005; Schmitt et al. 2010). The crucial role of eIF2 in the selection of initiation codons has been highlighted by the identification of mutant yeast that are capable of initiating translation on a non-AUG codon (Cigan et al. 1989; Dorris et al. 1995). Other studies have shown that a correct pairing between the initiation codon and the Met-tRNAt^{Met} anticodon induces the release of the e/aIF2 ribosome in a GDP-bound state. The nucleotide state of e/aIF2 is thus linked to the selection of the initiation codon (Huang et al. 1997).

At the very beginning of work on eIF2, sequence alignments had predicted structural homologies between its γ-subunit and EF1A, the factor responsible for binding aminoacylated tRNA elongators during the translation elongation stage (Erickson and Hannig 1996) (see next section). In the end, it was the determination of 3D structures of archaeal versions of aIF2 that enabled us to understand the structural organization of this heterotrimeric protein, as well as the binding mode of the methionylated initiator tRNA (Schmitt et al. 2002; Sokabe et al. 2006; Yatime et al. 2006; Stolboushkina et al. 2008; Schmitt et al. 2012). Consistent with sequence alignment predictions, the aIF2γ subunit contains three structural domains and strongly resembles EF1A (compare Figures 7.2(b) and 7.4(b), and see also the following paragraph). GTP-Mg^{2+} is bound to domain I. The conformations of two switch regions and the loop containing the GKT sequence control the factor's nucleotide status (Dubiez et al. 2015). aIF2α is composed of three domains (Figure 7.2). Its C-terminal domain is linked to domain II of aIF2γ. aIF2β has a long N-terminal helix, followed by a structural unit containing two subdomains, an α-β domain and a zinc-binding domain.

The N-terminal helix alone anchors β to γ (Figure 7.2). In the heterotrimer, the peripheral α and β subunits do not interact with one another. A comparison of all available structures also showed that γ, the C-terminal domain of α and the N-terminal helix of β, formed a rigid structural unit surrounded by two mobile wings (domains I and II of α and the central domain of β; Figure 7.2). The mobility of these two wings is important for the function of this factor.

In the structure of the TC (aIF2:GDPNP:Met-ARNt$_i^{Met}$), the acceptor stem of the methionylated initiator tRNA is bound to aIF2 via domain III of α and domains I and II of γ, while the aIF2β subunit does not appear to contribute to its binding. Unexpectedly, the binding mode of tRNA to aIF2γ is totally different from that observed for the binding of an aminoacylated elongator tRNA to EF1A, despite the very close structural similarity of these two proteins (compare Figures 7.2 and 7.4) (Nissen et al. 1995; Schmitt et al. 2012).

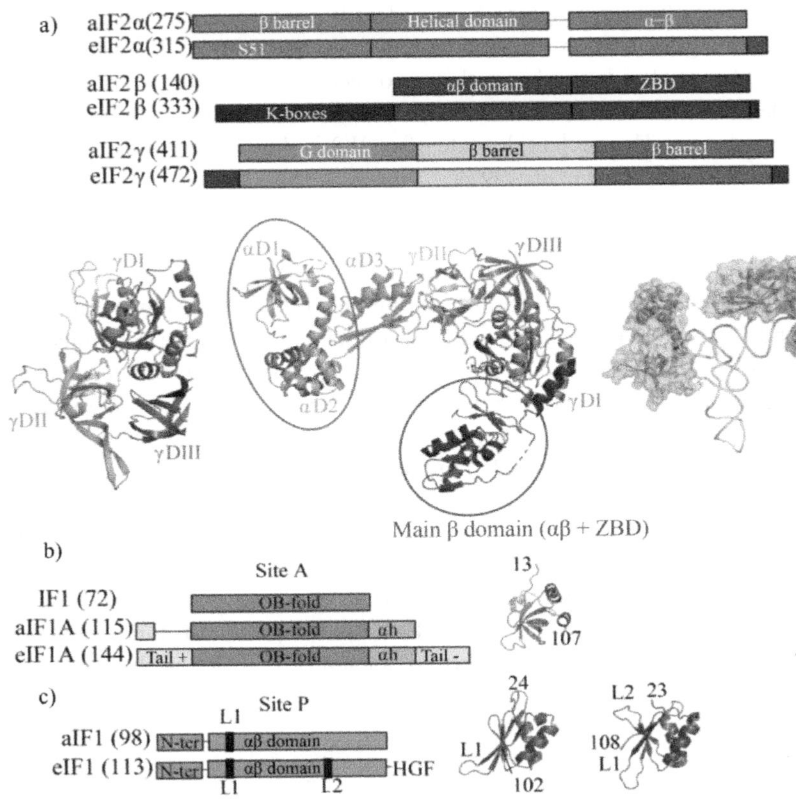

Figure 7.2. *Structural organization of common initiation factors in archaea and eukaryotes. For a color version of this figure, see www.iste.co.uk/clouet/archaea2.zip*

COMMENT ON FIGURE 7.2.– *(a) The three subunits of a/eIF2. Extensions specific to eukaryotic factors are in dark colors. Eukaryotic eIF2α have an acidic C-terminal extension. Eukaryotic eIF2β have an additional domain required for binding to eIF5 and eIF2Bε. eIF2γ have N-terminal and C-terminal extensions compared with archaeal versions. The structure of aIF2 γ from Sulfolobus solfataricus is shown in the bottom left of view A (PDB ID code 4RD4 (Dubiez et al. 2015)). GTP-binding domain I is in green, domain II is in light green and domain III in dark green. The same color code is used for the EF1A domains in Figure 7.3. The complete heterotrimeric aIF2 protein is shown in the middle view. The α subunit is colored in cyan, the β subunit is colored red and the γ subunit is colored in green. The two mobile wings of aIF2 corresponding to the αD12 domains and the central β domain are circled. (b) The structures of e/aIF1A and the bacterial homolog IF1. The structure of aIF1A from P. abyssi (PDB ID 4MNO) is shown in orange. The basic and acidic extensions required for long-distance scanning are represented by yellow boxes. The archaeal version of the factor contains a helical domain at the C-terminal end, but no acidic extension. A short N-terminal tail is present in aIF1A. The organization of the bacterial IF1 protein is also shown. As a result of their similarity, the IF1 and e/aIF1A proteins are considered universal initiation factors. (c) Structure of e/aIF1. The structure of aIF1 from Methanocaldococcus jannaschii (PDB ID 4MO0 (Coureux et al. 2016)) is shown in magenta, and that of eIF1 from S. cerevisiae (PDB ID 2OGH (Reibarkh et al. 2008)) is in purple. In all panels, residue numbers in brackets refer to P. abyssi proteins for archaea, to human protein for eukaryotes and to E. coli for IF1.*

7.1.2. *The 30S initiation complex in Pyrococcus abyssi*

The *Pyrococcus abyssi* translation initiation complex containing all of the players (30S:mRNA:TC:aIF1:aIF1A) was reconstituted in vitro and studied by cryogenic electron microscopy (cryo-EM) (Coureux et al. 2016). This complex contains a model mRNA derived from a natural *P. abyssi* mRNA possessing a nine-base complementary SD sequence of 16S rRNA. Analysis of the microscopy images revealed two initiation complex conformations (Figure 7.3).

In these two states, aIF1A occupies site A and aIF1 is located in front of site P. These positions correspond to those observed in eukaryotic complex structures (Lomakin et al. 2003; Weisser et al. 2013; Llacer et al. 2015).

In both states, the e/aIF2 factor contacts the ribosome at helix h44 via domain III of its γ subunit. Interestingly, this contact point has never been observed in eukaryotic initiation complex structures, although it had been suggested in these organisms by labeling experiments (Shin et al. 2011). In the major conformation, named IC0-P_{REMOTE}, the initiator tRNA bound to aIF2 is outside site P (Figure 7.3). The structure of the TC (aIF2:GDPNP:Met-$RNAt_i^{Met}$) is unconstrained, as it is similar to the crystallographic structure that has been defined in the absence of ribosome.

In the minor conformation, known as IC1-P_{IN}, the initiator tRNA anticodon is located at site P. aIF2 is still linked to the tRNA acceptor stem and to h44. Thus, the conformation of the TC changes from the unconstrained state observed in IC0-P_{REMOTE} to the constrained state observed in IC1-P_{IN}.

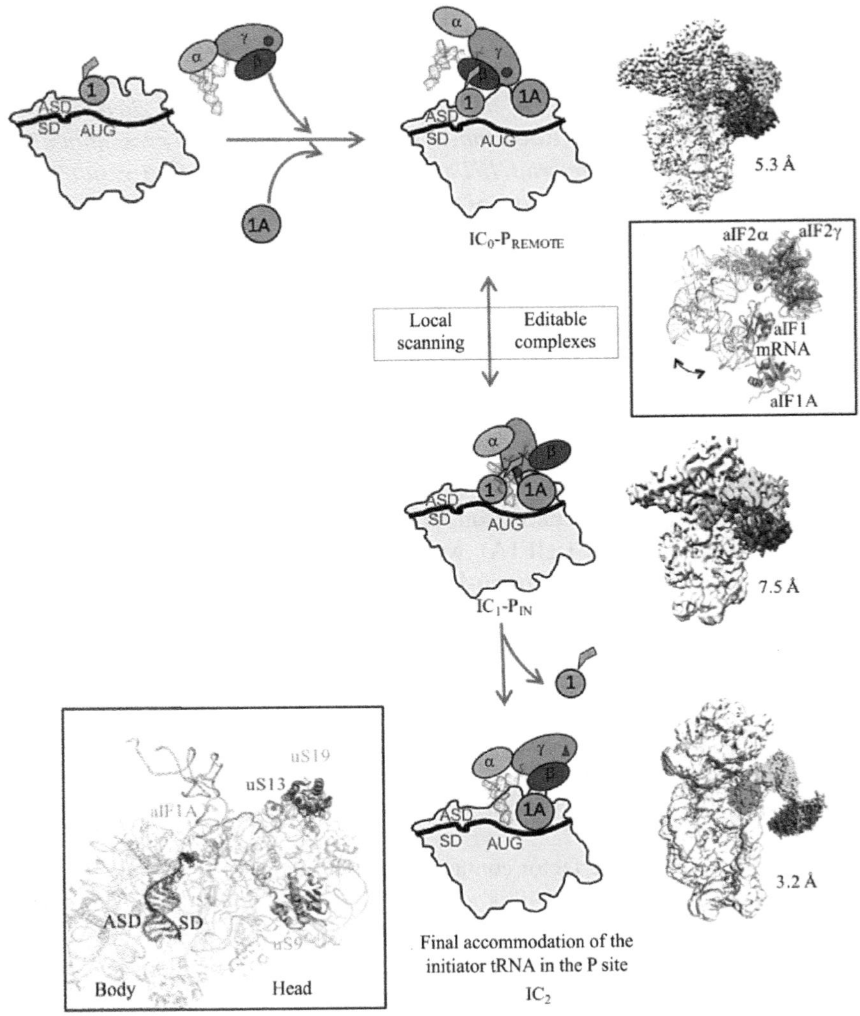

Figure 7.3. *Translation initiation stages in P. abyssi. For a color version of this figure, see www.iste.co.uk/clouet/archaea2.zip*

COMMENT ON FIGURE 7.3.– *Cryo-EM structures are from Coureux et al. (2016, 2020) (PDB ID codes: 5JB3, 5JBH, 6SWC). On the right of the figure, the positions of the ternary complex in IC_0-P_{REMOTE} and IC_1-P_{IN} are compared after superposition of the 30S "bodies". The color code is the same as in Figures 7.1 and 7.2, with light colors for IC_0-P_{REMOTE} and dark colors for IC_1-P_{IN}. For greater* *clarity, the mobile wings of aIF2 are omitted. In the insert at the top of the figure, the view shows the superposition of the two states, illustrating the oscillatory movement of tRNA that is only stabilized if a start codon is found at site P. Once pairing is achieved, the definitive positioning of the initiator tRNA at site P triggers the release of aIF1. The accommodated state of the initiator tRNA is observed in the IC2 complex. It is stabilized in part by interactions with the tails of three universal ribosomal proteins, uS9, uS13 and uS19 (see insert at the bottom of the figure).*

The deduced model is that, at the start of archaeal translation, tRNA bound to aIF2 oscillates between the two positions observed in IC0-P_{REMOTE} and IC1-P_{IN} (Figure 7.3). These two positions are in equilibrium, and the transition from one position to the other reflects the dynamics of the initiation complex, testing for the presence of a start codon in site P. The TC acts like a spring, pulling the initiator tRNA out of site P until its position is stabilized by codon-anticodon pairing. When a suitable codon is found, its pairing with the anticodon of the initiator tRNA stabilizes the IC1-P_{IN} conformation by compensating for the structural constraint on the TC. This lengthens the residence time of the initiator tRNA in site P and triggers the subsequent initiation stages. Indeed, because aIF1 and tRNA have partially superimposed binding sites, definitive tRNA accommodation at site P is incompatible with the presence of aIF1. The positioning of tRNA thus causes the factor to leave. Finally, the release of aIF1 also triggers the loosening of contact between aIF1 and aIF2. This contact disruption then triggers the release of the Pi group resulting from hydrolysis of the GTP bound to aIF2, making the process irreversible. This sequence of events following the recognition of the start codon is consistent with what has been observed in eukaryotes (Algire et al. 2005).

The dynamic nature of the aIF1-induced initiation complex, necessary for start codon selection, has been confirmed by ribosome toeprinting experiments (Monestier et al. 2018) (Figure 7.3) and by cryo-EM structure determination of an initiation complex lacking aIF1 at 3.2 Å resolution (Coureux et al. 2020). The stem and anticodon loop of the initiator tRNA are stabilized by a network of interactions involving the C-terminal tails of three universal ribosomal proteins and modified rRNA bases (Figure 7.3). Interestingly, this structure, supported by mapping experiments, revealed the presence of a large number of N4-acetylcytidines distributed throughout the rRNA16S structure, suggesting a role for these modifications in hyperthermostability. This result was subsequently confirmed in other *Thermococcales* (Sas-Chen et al. 2020). Other modified 16S rRNA bases play a functional role in initiation, as previously observed in eukaryotes and bacteria (Grosjean et al. 2008; Ferreira-Cerca 2017; Knüppel et al. 2021). These data also illustrate the key role of the ribosome itself in the selection of start codons.

7.1.3. *Comparison with eukaryotic eIF1, eIF1A and eIF2 factors*

As mentioned at the start of section 7.1, eukaryotes and archaea share a common structural core. Current data have indeed uncovered similarities for start codon selection in these two domains of life (positions and roles of e/aIF1; e/aIF1A, aIF2 (Hasenöhrl et al. 2006, 2009; Coureux et al. 2016, 2020)).

Remarkably, the process of long-distance mRNA scanning in eukaryotes involves peptide sequence elements that are specific to the eukaryotic versions of the three IF. Thus, eIF1A possesses N-terminal and C-terminal extensions that have been shown to be essential for mRNA scanning by the initiation complex (Mitchell and Lorsch 2008; Saini et al. 2010). Also, the L2 loop of the eIF1 factor exhibits a characteristic sequence involved in eukaryotic tRNA initiator recognition (Thakur and Hinnebusch 2018) (Figure 7.2). For eIF2, it appears that the two mobile wings have evolved differently in the two domains of life. The positioning of eIF2α domains I and II in site E, observed in cryo-EM structures of eukaryotic PICs (Llacer et al. 2015, 2018), is linked to the recognition of specific mRNA elements during scanning, such as the Kozak motif. eIF2α also has an acidic C-terminal extension, the function of which during initiation remains to be clarified (Naveau et al. 2013). With regard to eIF2β, the most striking difference concerns the presence of an additional domain in eukaryotes. This is involved in regulating the nucleotide state of the eIF2 factor. Indeed, starting from the minimal central complex, eukaryotes have evolved toward more sophisticated processes that have enabled the emergence of additional regulatory mechanisms, such as the integrated stress response (Harding et al. 2003; Pakos-Zebrucka et al. 2016). These mechanisms involve the regulation of GTP hydrolysis by eIF5 through an interaction with the eukaryotic-specific domain of eIF2β (Figure 7.3) and the regulation of GDP/GTP exchange on eIF2, which is controlled by the guanine nucleotide exchange factor B (eIF2B) (GEF) and the phosphorylation of a conserved serine in eIF2α. To date, no orthologs of eIF5 and B (eIF2B) catalytic subunits have been identified in archaea (Table 7.2).

7.1.4. *The different patterns of mRNA recruitment in archaea*

The evolution of translation initiation mechanisms is also linked to the evolution of mRNA recruitment, and therefore of the 5′ untranslated regions (5′UTR) of mRNA. For example, mRNAs with very short or even nonexistent 5′ UTRs, known as leaderless mRNAs, are present in all domains of life (Condo et al. 1999; Brenneis et al. 2007; Huber et al. 2019). Leaderless mRNAs are abundant in some archaea, such as *Sulfolobus solfataricus* (Condo et al. 1999) and *Haloferax volcanii* (Brenneis et al. 2007), and less abundant in others, such as *P. abyssi* (Toffano-Nioche et al. 2013) and *Aeropyrum pernix* (Torarinsson et al. 2005). The mechanism of translation initiation of leaderless mRNAs is still a subject of debate. In *S. solfataricus*, the formation of a stable complex between a leaderless mRNA and the 30S subunit has been shown to depend on the presence of the initiator tRNA (Torarinsson et al. 2005). However, the possibility of direct initiation with 70S ribosomes has yet to be studied in detail. The initiation of leaderless mRNA translation is best documented in bacteria (Grill et al. 2000; Moll et al. 2002, 2004; de Groot et al. 2014; Beck and Moll 2018; Sawyer and Cortes 2022). Although favorable IF2 action and unfavorable IF3 action have been observed, both pathways involving 30S or 70S are possible, and further studies are needed to determine which of these two pathways is more prevalent in vivo. Interestingly, it is possible that the evolution of messenger RNA recruitment patterns is linked to the evolution of ribosomes themselves. Indeed, ribosomal protein content varies in the different branches of archaea (Lecompte et al. 2002). The appearance or disappearance of certain proteins, such as eS26, could be correlated with the recruitment of messenger RNAs lacking SD sequencing. Indeed, in eukaryotes, eS26 is involved in the recognition of Kozak elements (Ferretti et al. 2017) and is observed interacting with the 3′ end of mRNAs in cryo-EM structures (Llacer et al. 2018).

7.1.5. *Assembly with the large subunit and formation of the elongation-competent ribosome*

In eukaryotes and archaea, the late stage of translation initiation occurring after e/aIF2:GDP departure involves the two factors, e/aIF1A and e/aIF5B, for final verification of the presence of an initiator tRNA and the large subunit junction (Choi et al. 1998; Pestova et al. 2000; Roll-Mecak et al. 2004) (Figure 7.1). A cryo-EM structure describing the function of aIF1A and aIF5B on the small archaeal subunit was very recently obtained (Kazan et al. 2022). This structure illustrates how the two proteins contribute to stabilizing the initiator tRNA in a conformation that is favorable for assembly. Equivalent findings have been obtained in eukaryotes (Brown et al. 2022; Lapointe et al. 2022). Strikingly, in bacteria, two orthologous factors exist, IF1 (homologous to e/aIF1A) and IF2 (homologous to e/aIF5B), and also function for the junction of the large ribosomal subunit (Rodnina 2018) (Table 7.2). This assembly step therefore has a universal nature. Furthermore, several studies have shown that in some cases, eukaryotic translation initiation uses eIF5B instead of eIF2, so as to ensure the recruitment of the initiator tRNA (Pestova et al. 2008; Terenin et al. 2008). This argues in favor of an ancestral translation initiation mechanism involving e/aIF5B/IF2 and e/aIF1A/IF1. This could be the mechanism used in LUCA (Last Universal Common Ancestor). Evolution could then have selected the formylation of the initiator tRNA in bacteria to favor the specificity of this step, whereas this enhancement would have been brought about in the eukaryotic/archaeal branch thanks to the emergence of e/aIF2.

7.2. Elongation peptide synthesis

The translation elongation step involves a succession of decoding, peptide bond formation and translocation cycles. The basic mechanism is very similar in bacteria and eukaryotes, which have been extensively studied (Dever et al. 2018; Rodnina 2018). Sequence and structural similarities, as well as biochemical data, suggest that the overall mechanism of elongation may also apply to the case of archaea (Woese et al. 1990; Dennis 1997; Vitagliano et al. 2001; Ito et al. 2014; Narrowe et al. 2018) (Figure 7.4(a); Table 7.3). The elongation steps are mainly controlled by the action of two GTPase-active proteins: aEF1A and aEF2 (Figures 7.4(b–d)). Many phylogenetic studies have been carried out using the sequences of these two proteins (Iwabe et al. 1989; Rivera and Lake 1992; Roger et al. 1999). In fact, these proteins are found in all three domains of life, and their activity is linked to a molecular process that appeared very early on in evolution. The sequences of these proteins have also been used to study phylogeny within the archaeal domain (Cammarano et al. 1992; Creti et al. 1994; Hashimoto and Hasegawa 1996; Brown and Doolittle 1997).

The trees obtained are consistent with the phylogenies deduced from the study of ribosomal RNAs, showing a bifurcation of the branches of the *Crenarchaeotes* and *Euryarchaeotes* within the *archaea* domain. The evolution of these proteins is directly linked to the evolution of ribosomes themselves (Hartman and Smith 2010). In fact, current data suggest that the ribosome began with the appearance of the large subunit with a peptide transfer center (PTC). EF1A is a monomeric protein made up of three structural domains (Figure 7.4(b)). From a molecular perspective, phylogenetic studies, aided by comparisons of 3D structures, suggest that the common ancestor of translational GTPases probably corresponds to a Ras-type GTPase (corresponding to domain I in Figure 7.4(b)), to which an OB-fold structural domain (corresponding to domain II in Figure 7.4(b)) has been added. This ancient elongation factor is thought to be the precursor of EF1A. EF2 would have appeared later, at the same time as the small ribosome subunit, from the evolution of the two domains found in EF1A (Figure 7.1(d)). Another consequence of the evolution of the EF1A ancestor is that it must have been connected to the evolution of transfer RNAs (Brown and Doolittle 1997). Consistent with this line of thought, the γ subunit of e/aIF2 would have arisen from the late evolution of EF1A and would have been specialized (thanks also to the addition of α and β subunits) in the recognition of the initiator tRNA.

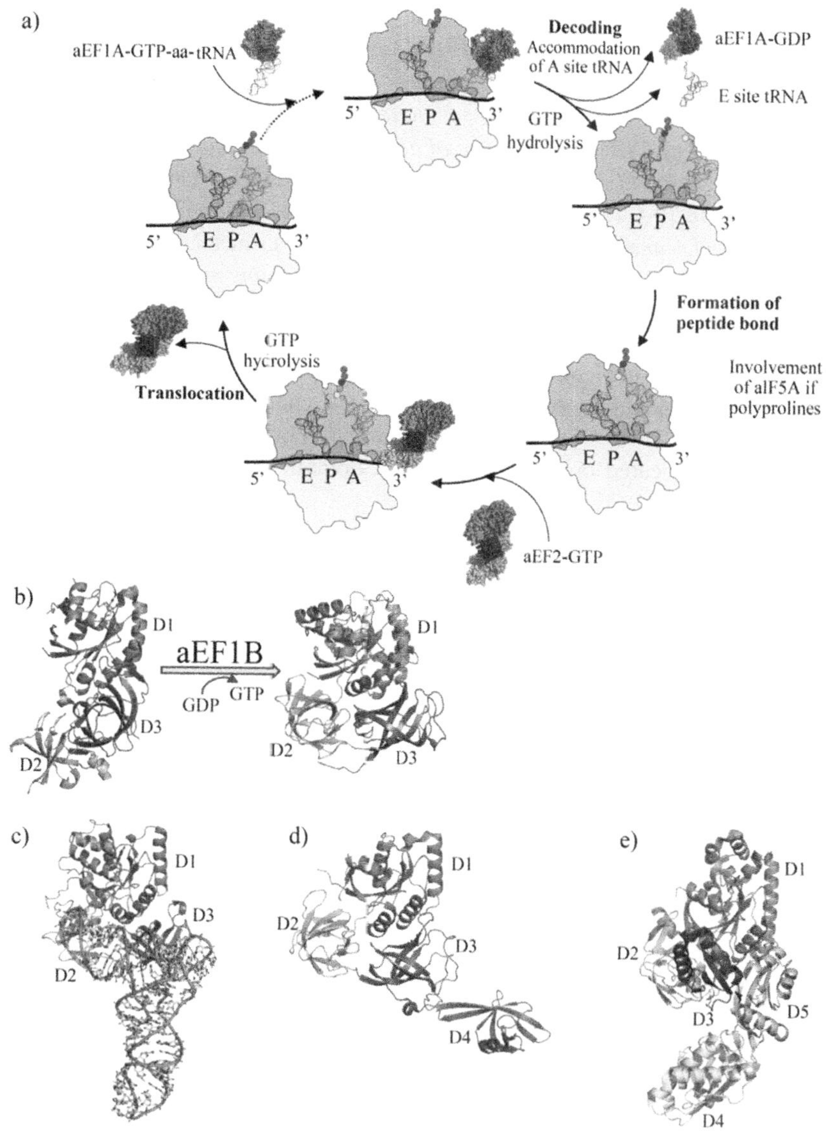

Figure 7.4. *Key stages and players in translation elongation. For a color version of this figure, see www.iste.co.uk/clouet/archaea2.zip*

COMMENT ON FIGURE 7.4.– *(a) Model of the elongation cycle of peptide synthesis in archaea. (b) Structure of aEF1A. Left view: aEF1A-GDP (3WYA (Ito et al. 2014)* *and 1JNY (Vitagliano et al. 2001)). Right view: EF1A-GTP (3WXM (Kobayashi et al. 2010) and 6JI2 (Maruyama et al. 2019)). Factor recycling is catalyzed by EF1B. (c) Structure of the bacterial ternary complex EF1A-GDPNP-Phe-tRNA^{Phe} (1TTT (Nissen et al. 1995)), which serves as a model for the binding of aminoacylated tRNA elongators to aEF1A. (d) Structure of aSelB from Methanococcus maripaludis (4ACA (Leibundgut et al. 2005)). D4 is involved in SECIS element recognition. (e) Structure of aEF2-GTP (Tanzawa et al. 2018). The D1 and D2 domains are conserved in aEF1A and aEF2. D3 is only encountered in aEF1A. aEF2 has three specific domains in red, yellow and blue. These three domains mimic the positioning of tRNA on aEF1A.*

Like all GTPases, nucleotide cycling on these proteins is controlled by two structural regions, known as switch regions, which change conformation (ON or OFF) according to GTP or GDP binding (Vetter and Wittinghofer 2001). These two switch regions are located on the D1 structural domain shared by all translational GTPases.

Bacteria	Eukaryotes	Archaea	Function
EF1A (EF-Tu)	eEF1A	aEF1A	Delivers tRNA elongator to site A
EF1B (EF-Ts)	eEF1Bαβ	aEF1Bα	EF1A GDP/GTP exchange factor
EF2 (EFG)	eEF2	aEF2	Accelerates translocation
EFP	eIF5A	aIF5A	Polyproline translation
L10/L12	P0/P1/P2	P0/P1	L12/P stalk
SRP54 (Ffh) ARN 4.5S	SRP54 SRP19 SRP14 SRP68 SRP72 SRP9 ARN 7S	SRP54 SRP19 ARN 7S	SRP (signal recognition particle)
SRα (FtsY)	SRα SRβ	SRα	SRP receptor (SR)

Table 7.3. *Players in translation elongation in the three domains of life*

7.2.1. Recruitment of aminoacylated tRNA to site A

GTP-bound aEF1A brings the aminoacylated elongating transfer RNAs to site A of the ribosome. The mode of tRNA binding to the archaeal protein is assumed to be close to that observed in bacteria (Figure 7.4(c)). Studies in bacteria and eukaryotes have shown that during the initial stages of selection, the EF1A:aa-tRNA complex is bound to the ribosome in a configuration that prevents premature interaction of the aminoacyl end of tRNA with the peptidyl transferase center (PTC) of the large subunit. The codon:anticodon interaction at the decoding center on the small subunit induces GTP hydrolysis on aEF1A. This is triggered by the interaction of the sarcin-ricin loop (SRL) of the large subunit, with the loop of the switch 2 region of aEF1A (Voorhees et al. 2010). aEF1A is then released from the ribosome in its GDP-bound form (Figure 7.1(b)), and the aminoacylated tRNA is released. Its aminoacylated end can then be accommodated in the PTC site. The rate of this step is critical. It is directly linked to the stability of the codon:anticodon interaction. Thus, the specificity of the decoding step would be achieved by a two-step mechanism, involving two successive verifications of codon:anticodon pairing at the decoding center (Shao et al. 2016; Rodnina 2018).

EF1A-GDP has a very different conformation to EF1A-GTP (Figure 7.4(b)). Indeed, the relative position of domains II and III to domain I is very different in the two nucleotide states. This shift accompanies the transition of the switch regions from the ON to the OFF state. EF1A-GDP is recycled into its active form, EF1A-GTP, which is capable of binding aminoacylated tRNA elongators through the action of the EF1B exchange protein. This protein is composed of two subunits in yeast, α and β. Subunit α is the catalytic subunit. It is the only one present in archaea. aEF1Bα is a shortened version of eEF1Bα. Its sequence is homologous to half the C-terminal domain of eEF1Bα. The structure of the complex between aEF1A and aEF1Bα has been resolved, and the role of aEF1Bα has been clarified (Suzuki et al. 2021).

7.2.2. The special case of the SelB factor

Selenium is a trace element that can be incorporated, in a co-translational way, in the form of selenocysteine into some very specific proteins in all three domains of life. In archaea, methanogens are known to utilize selenium (Romero et al. 2005; Atkinson et al. 2011; De Lise et al. 2021). Selenocysteine, still known as the 21st amino acid, is encoded by a UGA stop codon within the mRNA sequence and by a stem-loop structure located upstream of this codon (SECIS (selenocystein inserting sequence)). These two elements must be recognized for the UGA codon to be decoded into selenocysteine and no longer identified as a stop codon. The UGA codon in question is recognized by a specialized transfer RNA, $tRNA^{Sec}$, which possesses an additional anticodon. This $tRNA^{Sec}$ is supported by a specialized elongation factor, which differs from the classical aEF1A factor, the SelB factor (Commans and Böck 1999; Rother et al. 2000). This factor comprises three domains that are similar to the EF1A elongation factor, with a specific C-terminal domain involved in recognizing the SECIS element (Leibundgut et al. 2005).

Lastly, pyrrolysine (or the 22nd amino acid) is another unusual amino acid (Srinivasan et al. 2002; De Lise et al. 2021). It is a lysine derivative with a pyrroline ring attached to the lysine side chain. Its incorporation during protein synthesis requires a specialized tRNA and a UAG-type stop codon. Its distribution seemed narrower than that of selenocysteine, seemingly only being used in *Methanosarcina* and certain anaerobic bacteria. However, a recent analysis shows that pyrrolysine is more widespread (Brugère et al. 2018). The uptake of the pyrrolysin-tRNA is achieved through the canonical elongation factor EF1A (Atkinson et al. 2011; Yanagisawa et al. 2015).

7.2.3. *Translocation*

EF2, also known as translocase, catalyzes ribosome translocation, as well as the release of deacetylated transfer RNAs at site E (Figure 7.4(a)). The 3D structure of aEF2 from *Pyrococcus horikoshii* has been defined (Tanzawa et al. 2018). Despite this, very few studies concerning the translocation mechanism in archaea have been carried out, and the current state of knowledge has led to the generalization of studies carried out in eukaryotes and bacteria (for a review, see Achenbach and Nierhaus 2015; Ling and Ermolenko 2016; Dever et al. 2018; and Rodnina 2018).

After the departure of EF1A, the peptidyl part is transferred from the peptidyl-tRNA of site P to the aminoacyl-tRNA of site A via peptide bond formation. The reaction takes place at the PTC site of the large subunit. Thus, the nascent peptide, elongated by one amino acid, is attached to the tRNA site A, while a deacylated tRNA is localized at site P. EF2-GTP is then recruited (Figure 7.4(a)). EF2-GTP binds to the ribosome in a pre-translocation conformation, in which the tRNAs are in a hybrid A/P and P/E state (where the first letter indicates the position on the small subunit and the second letter indicates the position on the large subunit). In this conformation, the small ribosomal subunit is ratcheted in relation to the ribosome's large subunit. EF2 binding induces a conformational change in the small subunit (swiveling movement of the small subunit head relative to the body), which triggers translocation of the mRNA:2tRNAs complex and the transition of tRNAs to sites E and P. The translocation step is thus coupled to two ribosome movements: a ratcheting movement of the small subunit relative to the large one, and a swiveling movement of the head relative to the body within the small subunit (Figure 7.5). Translocation is achieved at the cost of hydrolysis of a GTP molecule, which is necessary for EF2 dissociation.

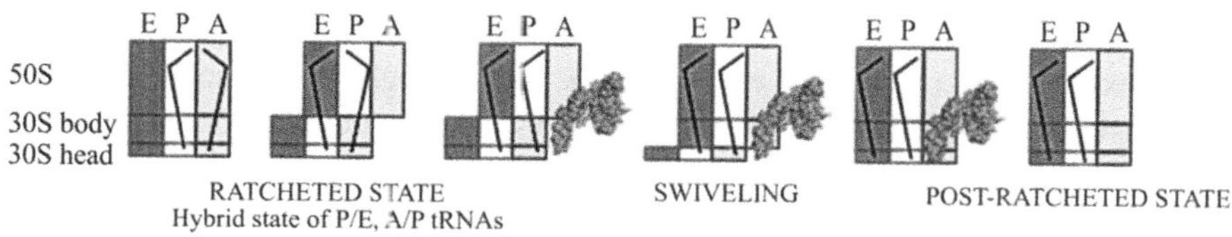

Figure 7.5. *Schematic representation of transfer RNA rearrangement during the translocation step. The figure was made from (Ling and Ermolenko 2016). The diagram shows the position of tRNAs relative to sites E, P and A of the large and small subunits. The hybrid A/P and P/E states are visible in views 2, 3 and 4, from the left. The EF2 factor is shown in green. For a color version of this figure, see www.iste.co.uk/clouet/archaea2.zip*

In both archaea and eukaryotes, EF2 carries a post-translational modification that is synthesized by a two-step mechanism (Su et al. 2013). First, a 3-amino-3-carboxypropyl group is added to a conserved histidine residue that is located at the end of domain 4 of EF2 (Figure 7.4(d)), then this intermediate (diphthine) is modified into a diphthamide residue. This modified amino acid is targeted by bacterial toxins, including the diphtheria toxin from which it takes its name. Phylogenetic studies show a complicated evolutionary history for aEF2 and the proteins involved in diphthamide synthesis. Phylogenetic trees are compatible with scenarios in which eukaryotes would have arisen within the Asgard group (Narrowe et al. 2018). At present, the biological function of diphthamide is not yet well understood. In eukaryotes, it appears to be involved in translation fidelity (Ortiz et al. 2006; Djumagulov et al. 2021).

7.2.4. *EF-P-e/aIF5A*

Peptide bond formation corresponds to a nucleophilic attack of the amino group of aa-tRNA on the carbon of the carbonyl group of the peptidyl-tRNA ester bond. This reaction is accelerated up to 10^7 by the ribosome, compared to the reaction in solution (Sievers et al. 2004). The ribosome catalyzes peptide bond formation with almost all amino acid combinations, without the help of auxiliary proteins. There is, however, one notable exception. During the synthesis of segments containing several prolines in a row, the ribosome pauses, because the rate of peptide bond formation becomes too slow. This ribosome pause is lifted by the recruitment of the EF-P protein in bacteria (Doerfel et al. 2013) or e/aIF5A in eukaryotes and archaea (Kim et al. 1998; Peat et al. 1998; Yao et al. 2003; Park et al. 2010; Gabel et al. 2013). This specialized factor binds to site E of the ribosome to promote a more favorable conformation of tRNAs at sites A and P, and of the nascent peptide in the exit tunnel (Melnikov et al. 2016). Interestingly, this factor carries a post-translational modification. This modification varies in the three domains of life. In *E. coli*, the modification corresponds to a lysyl-lysine. In eukaryotes, the factor has a lysine modified to hypusine. In aIF5A, the modification corresponds to hypusine or deoxyhypusine (Prunetti et al. 2016; Bassani et al. 2018, 2019). Current data show that this is a significant modification for factor function.

7.2.5. *GTPase recruitment platform*

The elongation mechanism and other translation steps depend on the recruitment and activation of GTPases (EF1A, EF2, e/aIF5B). The GTPase recruitment and activation center of translation is called the GTPase-associated center (GAC). The GAC contains two essential elements: an rRNA fragment called the SRL, which is necessary for activation, and the L12/P stalk, a protein-only extension involved in GTPase recruitment. These two elements cooperate to activate GTPases. The SRL was so named as it is the target of cytotoxins such as α-sarcin and ricin, which abolish translation (Grela et al. 2019). The L12/P stalk extends outside the ribosome. It is made up of several copies of acidic proteins. Unlike the SRL region, which is highly conserved in all three domains of life, the organization of this stalk has changed over the course of evolution. The P stalk of archaeal ribosomes is formed by the P0 ribosomal proteins in complex with three homodimers of the P1 protein (Maki et al. 2007; Naganuma et al. 2010; Ito et al. 2014) (Figure 7.6). Sequence alignments indicate that archaeal P1 proteins are similar to eukaryotic P1/P2 proteins, but distant from the bacterial analog bL12 (Grela et al. 2008; Choi et al. 2015).

Determination of the 3D structures confirmed these discrepancies. It was also shown that the C-terminal parts (CTDs) of eukaryotic P1/P2 and L12 proteins were responsible for binding elongation factors, even though their structures were very different. In archaea, it is the C-terminal tails of aP1 that interact with the aEF1A, aEF2 and aIF5B proteins (Mochizuki et al. 2012; Nomura et al. 2012). These interactions are thought to be important in directing elongation factors to the SRL when hydrolysis becomes necessary. Thus, even if the structures of these proteins are different in the three domains of life, they would play the same role and would therefore be three structural solutions that ensure the same function.

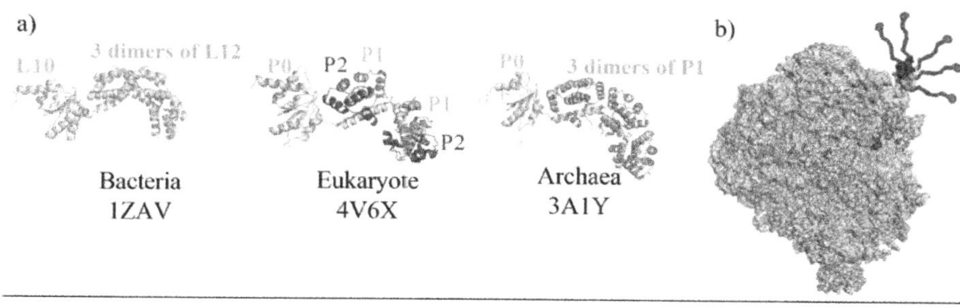

Figure 7.6. *GTPase recruitment platform or L12/P stalk. For a color version of this figure, see www.iste.co.uk/clouet/archaea2.zip*

COMMENT ON FIGURE 7.6.– *(a) Structural organization of the ribosome-anchored L12/P stalk region in the three domains of life; (b) structural model of the eukaryotic P stalk, drawn using PDB structure 4V88 (Ben-Shem et al. 2011). The ribosome-anchored part of the stalk is colored the same as in A. The flexible C-terminal part of the P1 and P2 proteins involved in GTPase recruitment is shown in blue. The sarcin-ricin loop is colored in red. The figure was adopted from Choi et al. (2015).*

7.2.6. *Localized translation at the membrane*

One-third of proteins are produced by membrane-associated ribosomes. The synthesis of these proteins is localized by the signal recognition particle (SPR) complex and its membrane-associated receptor (SR). SRP binds to the signal sequences of nascent proteins as they emerge from the ribosome exit tunnel. The SRP:peptide signal complex then binds to the membrane receptor. SRP organization varies in the three domains of life. Archaea show a hybrid organization of SRP compared with bacteria and eukaryotes (Gribaldo and Cammarano 1998; Eichler and Moll 2001; Egea et al. 2005; Zwieb and Bhuiyan 2010) (Table 7.1). Archaea resemble eukaryotes, with an SRP complex that contains a small 7S-type RNA and two associated proteins (SRP54 and SRP19). However, when it comes to the SR and the SRP receptor, archaea are more closely related to bacteria, as no homolog of the receptor's β subunit found in eukaryotes has been identified in archaea. Advances in our knowledge of SRP function have benefited from structural studies of archaeal representatives. Despite this, knowledge of this addressing system in archaea remains very limited. The development of genetic and cell biology tools for the structural studies of large assemblies by cryo-EM should make up for these shortcomings (Gupta et al. 2017).

7.3. Translation termination and ribosome recycling

When one of the three stop codons (UAA, UAG or UGA) reaches site A, a series of events takes place to hydrolyze the peptidyl-tRNA present at site P, releasing the polypeptide. This step is known as "termination".

Universally, termination requires the intervention of two so-called release factors (RF). The first, class 1, detects the presence of the stop codon and triggers hydrolysis to release the neosynthesized polypeptide. It is assisted by a factor, known as class 2, which is generally a G protein, and therefore capable of hydrolyzing GTP.

After translation has been completed, the ribosome must be prepared for a new translation cycle. We will only briefly mention the case of bacteria where, after polypeptide release, class 1 and 2 termination factors leave the ribosome. Two factors, RRF (ribosome release factor) and the GTP-bound translocase EF-G, then intervene to release the mRNA and the last tRNA at site P, as well as to dissociate the ribosome into small and large subunits. The initiation factor IF3 binds to the small subunit, preventing unintended reassociation of ribosomal subunits. In eukaryotes and archaea, termination and recycling are steps that are paired. Indeed, the ABCE1 protein, which plays a crucial role in the disassociation step, is also involved in the final termination step.

Bacteria	Eukaryotes	Archaea	Function
RF1 and RF2	eRF1	aRF1	Recognizes stop codons. RF1/2 and e/aRF1 have very little in common
RF3 (Tu-like)	eRF3 (Tu-like)	aEF1A	Class 2 termination factor, G protein. Assists class 1 factor
RRF	–	–	Subunit dissociation
EF-G	–	–	Assists RRF
–	ABCE1	ABCE1	Subunit dissociation. Intervenes after class 2 RF has left, but e/aRF1 is still present. Stimulates e/aRF1 activity

Table 7.4. *Players in translation termination and ribosome recycling in the three domains of life*

7.3.1. *The class 1 factor, aRF1*

In bacteria, two class 1 factors, RF1 and RF2, are required to recognize the three stop codons. RF1 recognizes UAA and UAG, while RF2 recognizes UAA and UGA. GTP binding and hydrolysis by the class 2 factor, RF3, are required for factor recycling after termination, but not for polypeptide chain release (Zavialov et al. 2001; Graf et al. 2018).

In archaea and eukaryotes, there is only one class 1 factor that recognizes all three codons (Frolova et al. 1994; Dontsova et al. 2000; Hellen 2018). The aRF1 factor closely resembles its eukaryotic counterpart eRF1. In contrast, aRF1 and eRF1 are distinctly different from the bacterial factors RF1 and RF2. The only common feature of all class 1 factors is the presence of a GGQ motif which, as we shall see later, is involved in peptidyl-tRNA hydrolysis. The glutamine in the conserved GGQ motif is methylated to N5 in bacteria and eukaryotes. The corresponding methylase, Mtq2, and its accessory protein, Trm112, are conserved in archaea, indicating that glutamine methylation is universal (Bourgeois et al. 2017). In the bacterial case, post-translational modification has been shown to facilitate peptide release by making the rate of hydrolysis less dependent on the C-terminal amino acid (Pierson et al. 2016). However, its role has yet to be precisely described in archaeal and eukaryotic cases. a/eRF1 is composed of three domains: domain I, or N, which carries a conserved sequence close to NIKS, domain II, or M, which carries the GGQ sequence, and a domain III, or C, which is involved in partner protein recognition.

7.3.2. *The class 2 archaeal factor, aEF1A*

The eukaryotic class 2 factor, eRF3, is an EF1A-like GTPase. In this case, GTP hydrolysis is required for the release of the neosynthesized polypeptide. Remarkably, in archaea, the class 2 factor does not exist, but its role is played by aEF1A. Furthermore, the 3D structure of the aEF1A:GTP:aRF1 complex bears an obvious resemblance to that of the "universal" TC EF-Tu:GTP:aa-tRNA (Figure 7.6). Thus, aEF1A can be seen as a vessel for transporting eRF1 to the ribosome, in the same way that e/aEF1A (or EF-Tu) transports aminoacylated tRNAs for polypeptide chain elongation. The precise mechanisms of translation termination have not been studied much in archaea, so we can only speculate based on the strong resemblance between eukaryotic and archaeal factors. aRF1, like eRF1, would act by mimicking a tRNA, to some extent. aEF1A would position aRF1 such that a domain I pocket, containing the conserved NIKS sequence, is placed close to the codon at the ribosome site A. As observed for eRF1 (Brown et al. 2015), this aRF1 domain could thus recognize the presence of one of the three stop codons in site A, much like the anticodon of a tRNA would. Indeed, when comparing the aEF1A:GTP:aa-tRNA and aEF1A:GTP:aRF1 complexes, the aRF1 domain I pocket is in the same position as the tRNA anticodon. In the eukaryotic case, which has been studied in more detail, great similarities are observed when comparing the accommodation of eEF1A:GTP:tRNA, after reading a sense codon, with that of eRF3:GTP:eRF1, after reading a stop codon (Shao et al. 2016).

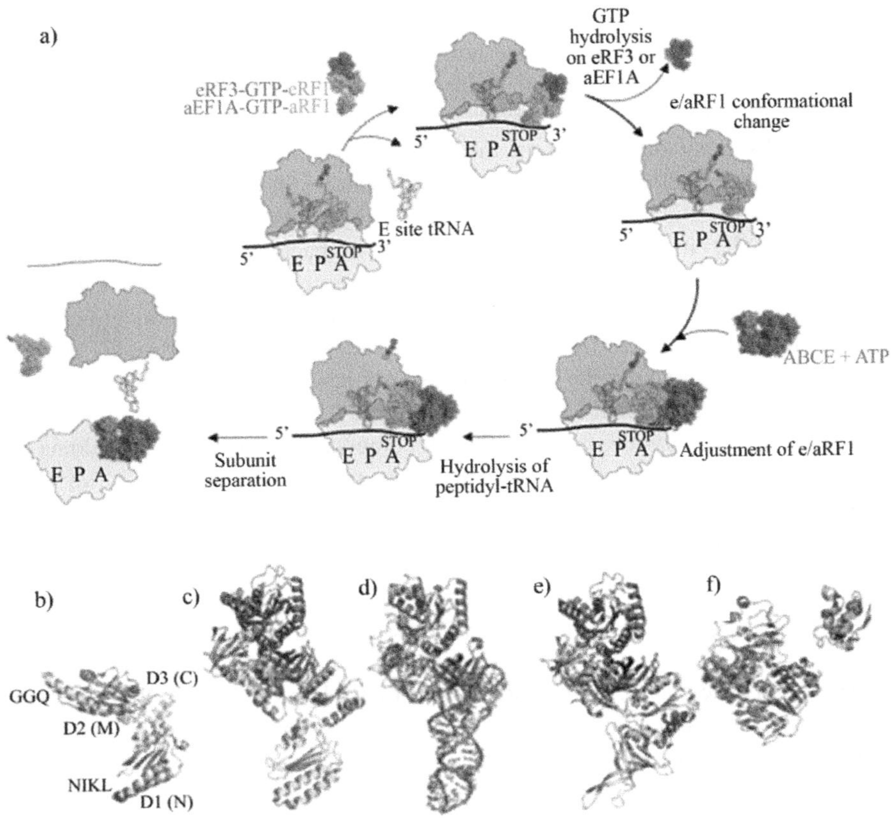

Figure 7.7. *Key stages and players in translation termination and mRNA recycling. For a color version of this figure, see www.iste.co.uk/clouet/archaea2.zip*

COMMENT ON FIGURE 7.7.– *(a) Model of the termination steps of peptide synthesis in archaea. Within aRF1 (in yellow), the GGQ motif region is indicated in blue and that of the NIKS motif in orange. (b) Structure of aRF1 from Aeropyrum pernix (3AGK (Saito et al. 2010)). The GGQ motif region is shown in blue, and that of the NIKS motif (NIKL, in this case) in orange. GGQ residues are not visible on the* *structure. (c) Structure of the aEF1A-GDPNP-aRF1 complex (3VMF (Kobayashi et al. 2010)). Color codes for aEF1A are the same as in Figure 7.1(b). (d) Structure of the bacterial ternary complex EF1A-GDPNP-Phe-tRNA^{Phe} (1TTT (Nissen et al. 1995) (see Figure 7.1(c))) shown for comparison with panel C. (e) Structure of the ternary complex aEF1A-GTP-aPelota from A. pernix (6JI2 according to Maruyama et al. (2019)). Domains are colored with the same code as for panel C, substituting aPelota to aRF1. (f) Structure of ABCE1 from S. solfataricus (3OZX). The N-terminal domain is not visible. That of P. abyssi (5LW7) is shown in blue.*

As described above (see section 7.2), the SRL activates GTP hydrolysis by eEF1A and the aminoacylated end of the tRNA makes a major move to position itself close to the PTC. In the case of termination, the SRL activates GTP hydrolysis by eRF3, and the M and C domains of eRF1 make the same type of movement to position themselves close to the PTC, after the departure of eRF3-GDP. This conformational change brings the GGQ motif to the same position as the tRNA acceptor end, in a good position to activate the ribosome's PTC, so as to catalyze hydrolysis of the peptidyl-tRNA ester bond at site P. As we shall see later, peptidyl-tRNA hydrolysis will be further stimulated upon recruitment of ABCE1. It is worth noting that the precise mechanisms that enable GTPase activation and accommodation differ when comparing elongation and termination, arguably improving the specificity of the system (Shao et al. 2016; Hellen 2018). The fact that aEF1A replaces eRF3 in archaea further supports the idea of unity in ribosome function during elongation and termination phases in eukaryotes and archaea. It will be worth deciphering the details of the termination mechanisms in archaea in order to determine whether the substitution of eRF3 by aEF1A makes termination and elongation even closer to one another in archaea than they are in eukaryotes.

7.3.3. *The ABCE1 factor*

The subunit dissociation step is achieved through the recruitment of the ABCE1 protein. This is a highly conserved protein in eukaryotes and archaea (Barthelme et al. 2011). This protein, belonging to the large family of ABC (ATP binding cassette) proteins, is capable of binding two ATP molecules in a "head-to-tail" conformation. ABCE1 also has an N-terminal domain carrying two $[4Fe\text{-}4S]^{2+}$ clusters (Barthelme et al. 2007). This ABCE1-dependent recycling process has only recently begun to be studied. Biochemical and structural studies have been carried out, in some cases using eukaryotic models, in others using archaeal models, and the data currently available show that ABCE1 functions very similarly in both domains (Becker et al. 2012). ABCE1 binds in the space between the two subunits of the ribosome, and its presence is incompatible with that of the translational GTPases e/aEF1A, e/aEF2 and eRF3. The ABCE1 protein therefore intervenes after the departure of the class 2 termination factor.

All available results, including those from structural studies of complexes, reflecting distinct steps in the subunit dissociation process (Becker et al. 2012; Kiosze-Becker et al. 2016; Heuer et al. 2017), have led to the following model (Gerovac and Tampe 2019; Gouridis et al. 2019). The N-terminal domain bearing the Fe-S group interacts with the M domain of e/aRF1. This leads to small adjustments in e/aRF1, which probably explain why ABCE1 is able in vitro to stimulate eRF1 and hydrolyze a peptidyl-tRNA (Shoemaker and Green 2011). This is the most favorable time for peptide release. ABCE1 undergoes a major conformational change, rotating its N-terminal domain by 150 degrees. This rotation enables the domain to repel the e/aRF1 factor, which then acts as a wedge to separate the two subunits. The binding of two ATP molecules by ABCE1 is essential for this movement to occur. The protein's conformational change is linked to a complex dynamic equilibrium involving three conformational states (open, intermediate and closed) for each of the two nucleotide sites. In this model, ATP hydrolysis is only required for factor release, not for dissociation activity (Gouridis et al. 2019; Nurenberg-Goloub et al. 2020). It should be stressed, however, that a possible role for ATP hydrolysis in dissociation activity cannot be ruled out. Indeed, several studies suggest that ATP hydrolysis could produce the chemomechanical force enabling subunit separation (Pisarev et al. 2010; Shoemaker and Green 2011; Becker et al. 2012). This point of debate is still actively being researched.

It is therefore clear that in eukaryotes and archaea, the termination and recycling steps are coupled, since the presence of eRF1 is necessary for the recruitment and action of ABCE1 which, in turn, stimulates the activity of eRF1 (Shoemaker and Green 2011).

7.3.4. *The special case of abnormal terminations*

Another interesting feature of archaea is their mRNA quality control mechanisms. There are numerous systems for releasing ribosomes trapped on defective mRNAs, and then degrading the corresponding mRNAs. No-go decay (NGD) and non-stop decay (NSD) are mechanisms for recognizing ribosomes blocked on an mRNA. For example, due to a secondary structure, a deteriorated nucleotide (NGD) or a truncated messenger upstream of the stop codon (NSD) (Shoemaker and Green 2012). In these cases, in eukaryotes, the stalled ribosome is recognized by a complex of two proteins, Pelota (also called Dom34 in

S. cerevisiae) and Hbs1 (Tsuboi et al. 2012). Pelota resembles eRF1 (Graille et al. 2008) and Hbs1 resembles aEF1A, so that the Pelota:Hbs1 complex resembles the eRF1:eRF3 complex (Chen et al. 2010). Indeed, the Pelota:Hbs1 complex binds to site A of the ribosome, and this binding marks the beginning of the ribosome dissociation process (Shoemaker et al. 2010; Becker et al. 2011). However, unlike e/aRF1, Pelota has neither the GGQ nor the NIKS sequence, and therefore does not act by recognizing a stop codon at site A and hydrolyzing peptidyl-tRNA. In archaea, we find the same system for releasing blocked ribosomes, with an aPelota protein orthologous to ePelota. Remarkably, however, in this case, aEF1A plays the role of Hbs1 (Kobayashi et al. 2010; Saito et al. 2010). Thus, in archaea, aEF1A plays three roles in (i) polypeptide chain elongation, (ii) termination and (iii) quality control. In eukaryotes, three paralogous proteins are specialized for each of these roles: eEF1A, eRF3 and Hbs1.

7.3.5. *Ribosome recycling after abnormal termination*

It is worth mentioning that ABCE1 is not only able to recycle ribosomes after canonical termination by aRF1/aEF1A (or eRF1/eRF3), but also after premature termination, following the action of aPelota/aEF1A (or Pelota/Hbs1). The ABCE1:Pelota interaction is quite similar to the ABCE1:aRF1 interaction (Becker et al. 2012). However, a major difference is that, lacking the GGQ motif, Pelota cannot release the peptide under synthesis. The ABCE1/Pelota pair can still dissociate ribosomal subunits in the presence of peptidyl-tRNA at site P. In contrast, at least in the eukaryotic case, prior hydrolysis of the peptide is required for the ABCE1/eRF1 pair to effectively dissociate subunits (Pisarev et al. 2010).

7.4. Conclusion

In this chapter, we have reviewed what we know about the different stages of translation in archaea. Many points still require further research. Among them, the mechanisms of mRNA translation without a 5′UTR region need to be deciphered. Moreover, even if the elongation stage is probably the most conserved stage in the three domains of life, biochemical and structural data concerning archaeal systems are still very patchy. Similarly, the termination stage is poorly documented for archaeal systems. New methodologies and the use of integrative approaches will undoubtedly provide essential new data for understanding archaea at the molecular level, and for clarifying the phylogenetic links with the other two domains of life.

7.5. References

Achenbach, J. and Nierhaus, K.H. (2015). The mechanics of ribosomal translocation. *Biochimie*, 114, 80–89.

Algire, M.A., Maag, D., Lorsch, J.R. (2005). Pi release from eIF2, not GTP hydrolysis, is the step controlled by start-site selection during eukaryotic translation initiation. *Mol. Cell.*, 20(2), 251–262.

Armache, J.P., Anger, A.M., Marquez, V., Franckenberg, S., Frohlich, T., Villa, E., Berninghausen, O., Thomm, M., Arnold, G.J., Beckmann, R. et al. (2013). Promiscuous behaviour of archaeal ribosomal proteins: Implications for eukaryotic ribosome evolution. *Nucleic Acids Res.*, 41(2), 1284–1293.

Atkinson, G.C., Hauryliuk, V., Tenson, T. (2011). An ancient family of SelB elongation factor-like proteins with a broad but disjunct distribution across archaea. *BMC Evol. Biol.*, 11, 22.

Ban, N., Nissen, P., Hansen, J., Moore, P.B., Steitz, T.A. (2000). The complete atomic structure of the large ribosomal subunit at 2.4 A resolution. *Science*, 289(5481), 905–920.

Ban, N., Beckmann, R., Cate, J.H., Dinman, J.D., Dragon, F., Ellis, S.R., Lafontaine, D.L., Lindahl, L., Liljas, A., Lipton, J.M. et al. (2014). A new system for naming ribosomal proteins. *Curr. Opin. Struct. Biol.*, 24, 165–169.

Barthelme, D., Scheele, U., Dinkelaker, S., Janoschka, A., Macmillan, F., Albers, S.V., Driessen, A.J., Stagni, M.S., Bill, E., Meyer-Klaucke, W. et al. (2007). Structural organization of essential iron-sulfur clusters in the evolutionarily highly conserved ATP-binding cassette protein ABCE1. *J. Biol. Chem.*, 282(19), 14598–14607.

Barthelme, D., Dinkelaker, S., Albers, S.V., Londei, P., Ermler, U., Tampe, R. (2011). Ribosome recycling depends on a mechanistic link between the FeS cluster domain and a conformational switch of the twin-ATPase ABCE1. *Proc. Natl. Acad. Sci. USA*, 108(8), 3228–3233.

Bassani, F., Romagnoli, A., Cacciamani, T., Amici, A., Benelli, D., Londei, P., Martens, B., Blasi, U., La Teana, A. (2018). Modification of translation factor aIF5A from *Sulfolobus solfataricus*. *Extremophiles*, 22(5), 769–780.

Bassani, F., Zink, I.A., Pribasnig, T., Wolfinger, M.T., Romagnoli, A., Resch, A., Schleper, C., Bläsi, U., La Teana, A. (2019). Indications for a moonlighting function of translation factor aIF5A in the crenarchaeum Sulfolobus solfataricus. *RNA Biology*, 16(5), 675–685.

Beck, H.J. and Moll, I. (2019). Leaderless mRNAs in the spotlight: Ancient but not outdated! *Microbiol. Spectr.*, 6(4). doi: 10.1128/microbiolspec.RWR-0016-2017.

Becker, T., Armache, J.P., Jarasch, A., Anger, A.M., Villa, E., Sieber, H., Motaal, B.A., Mielke, T., Berninghausen, O., Beckmann, R. (2011). Structure of the no-go mRNA decay complex Dom34-Hbs1 bound to a stalled 80S ribosome. *Nat. Struct. Mol. Biol.*, 18, 715–720.

Becker, T., Franckenberg, S., Wickles, S., Shoemaker, C.J., Anger, A.M., Armache, J.P., Sieber, H., Ungewickell, C., Berninghausen, O., Daberkow, I. et al. (2012). Structural basis of highly conserved ribosome recycling in eukaryotes and archaea. *Nature*, 482(7386), 501–506.

Ben-Shem, A., Garreau de Loubresse, N., Melnikov, S., Jenner, L., Yusupova, G., Yusupov, M. (2011). The structure of the eukaryotic ribosome at 3.0 Å resolution. *Science*, 334(6062), 1524–1529.

Benelli, D. and Londei, P. (2011). Translation initiation in Archaea: Conserved and domain-specific features. *Biochem. Soc. Trans.*, 39(1), 89–93.

Blombach, F., Matelska, D., Fouqueau, T., Cackett, G., Werner, F. (2019). Key concepts and challenges in archaeal transcription. *J. Mol. Biol.*, 431(20), 4184–4201.

Borck, G., Shin, B.S., Stiller, B., Mimouni-Bloch, A., Thiele, H., Kim, J.R., Thakur, M., Skinner, C., Aschenbach, L., Smirin-Yosef, P. et al. (2012). eIF2γ mutation that disrupts eIF2 complex integrity links intellectual disability to impaired translation initiation. *Mol. Cell.*, 48(4), 641–646.

Bourgeois, G., Letoquart, J., van Tran, N., Graille, M. (2017). Trm112, a protein activator of methyltransferases modifying actors of the eukaryotic translational apparatus. *Biomolecules*, 7(1), 7.

Brenneis, M., Hering, O., Lange, C., Soppa, J. (2007). Experimental characterization of cis-acting elements important for translation and transcription in halophilic archaea. *PLoS Genet*, 3(12), e229.

Brown, J.R. and Doolittle, W.F. (1997). Archaea and the prokaryote-to-eukaryote transition. *Microbiol. Mol. Biol. Rev.*, 61(4), 456–502.

Brown, A., Shao, S., Murray, J., Hegde, R.S., Ramakrishnan, V. (2015). Structural basis for stop codon recognition in eukaryotes. *Nature*, 524, 493–496.

Brown, Z.P., Abaeva, I.S., De, S., Hellen, C.U.T., Pestova, T.V., Frank, J. (2022). Molecular architecture of 40S translation initiation complexes on the hepatitis C virus IRES. *Embo J.*, 41(16), e110581.

Brugère, J.-F., Atkins, J.F., O'Toole, P.W., Borrel, G. (2018). Pyrrolysine in archaea: A 22nd amino acid encoded through a genetic code expansion. *Emerging Topics in Life Sciences*, 2(4), 607–618.

Cammarano, P., Palm, P., Creti, R., Ceccarelli, E., Sanangelantoni, A.M., Tiboni, O. (1992). Early evolutionary relationships among known life forms inferred from elongation factor EF-2/EF-G sequences: Phylogenetic coherence and structure of the archaeal domain. *J. Mol. Evol.*, 34(5), 396–405.

Chen, L., Muhlrad, D., Hauryliuk, V., Cheng, Z., Lim, M.K., Shyp, V., Parker, R., Song, H. (2010). Structure of the Dom34-Hbs1 complex and implications for no-go decay. *Nat Struct. Mol. Biol.*, 17, 1233–1240.

Choi, S.K., Lee, J.H., Zoll, W.L., Merrick, W.C., Dever, T.E. (1998). Promotion of met-tRNAiMet binding to ribosomes by yIF2, a bacterial IF2 homolog in yeast. *Science*, 280(5370), 1757–1760.

Choi, A.K., Wong, E.C., Lee, K.M., Wong, K.B. (2015). Structures of eukaryotic ribosomal stalk proteins and its complex with trichosanthin, and their implications in recruiting ribosome-inactivating proteins to the ribosomes. *Toxins (Basel)*, 7(3), 638–647.

Cigan, A.M., Pabich, E.K., Feng, L., Donahue, T.F. (1989). Yeast translation initiation suppressor sui2 encodes the alpha subunit of eukaryotic initiation factor 2 and shares sequence identity with the human alpha subunit. *Proc. Natl. Acad. Sci. USA*, 86(8), 2784–2788.

Commans, S. and Böck, A. (1999). Selenocysteine inserting tRNAs: An overview. *FEMS Microbiol. Rev.*, 23(3), 335–351.

Condo, I., Ciammaruconi, A., Benelli, D., Ruggero, D., Londei, P. (1999). Cis-acting signals controlling translational initiation in the thermophilic archaeon Sulfolobus solfataricus. *Mol. Microbiol.*, 34(2), 377–384.

Coureux, P.D., Lazennec-Schurdevin, C., Monestier, A., Larquet, E., Cladiere, L., Klaholz, B.P., Schmitt, E., Mechulam, Y. (2016). Cryo-EM study of start codon selection during archaeal translation initiation. *Nat. Commun.*, 7, 13366.

Coureux, P.D., Lazennec-Schurdevin, C., Bourcier, S., Mechulam, Y., Schmitt, E. (2020). Cryo-EM study of an archaeal 30S initiation complex gives insights into evolution of translation initiation. *Commun. Biol.*, 3, 58.

Creti, R., Ceccarelli, E., Bocchetta, M., Sanangelantoni, A.M., Tiboni, O., Palm, P., Cammarano, P. (1994). Evolution of translational elongation factor (EF) sequences: Reliability of global phylogenies inferred from EF-1 alpha(Tu) and EF-2(G) proteins. *Proc. Natl. Acad. Sci. USA*, 91, 3255–3259.

De Lise, F., Strazzulli, A., Iacono, R., Curci, N., Di Fenza, M., Maurelli, L., Moracci, M., Cobucci-Ponzano, B. (2021). Programmed deviations of ribosomes from standard decoding in Archaea. *Front Microbiol.*, 12, 688061.

Demo, G., Rasouly, A., Vasilyev, N., Svetlov, V., Loveland, A.B., Diaz-Avalos, R., Grigorieff, N., Nudler, E., Korostelev, A.A. (2017). Structure of RNA polymerase bound to ribosomal 30S subunit. *Elife*, 6, e28560.

Dennis, P.P. (1997). Ancient ciphers: Translation in Archaea. *Cell*, 89(7), 1007–1010.

Dever, T.E., Dinman, J.D., Green, R. (2018). Translation elongation and recoding in eukaryotes. *Cold Spring Harb Perspect. Biol.*, 10(8), a032649.

Djumagulov, M., Demeshkina, N., Jenner, L., Rozov, A., Yusupov, M., Yusupova, G. (2021). Accuracy mechanism of eukaryotic ribosome translocation. *Nature*, 600, 543–546.

Dmitriev, S.E., Stolboushkina, E.A., Terenin, I.M., Andreev, D.E., Garber, M.B., Shatsky, I.N. (2011). Archaeal translation initiation factor aIF2 can substitute for eukaryotic eIF2 in ribosomal scanning during mammalian 48S complex formation. *J. Mol. Biol.*, 413(1), 106–114.

Doerfel, L.K., Wohlgemuth, I., Kothe, C., Peske, F., Urlaub, H., Rodnina, M.V. (2013). EF-P is essential for rapid synthesis of proteins containing consecutive proline residues. *Science*, 339(6115), 85–88.

Dontsova, M., Frolova, L., Vassilieva, J., Piendl, W., Kisselev, L., Garber, M. (2000). Translation termination factor aRF1 from the archaeon *Methanococcus jannaschii* is active with eukaryotic ribosomes. *FEBS Lett.*, 472(2–3), 213–216.

Dorris, D.R., Erickson, F.L., Hannig, E.M. (1995). Mutations in GCD11, the structural gene for eIF-2 gamma in yeast, alter translational regulation of GCN4 and the selection of the start site for protein synthesis. *EMBO J.*, 14(10), 2239–2249.

Dubiez, E., Aleksandrov, A., Lazennec-Schurdevin, C., Mechulam, Y., Schmitt, E. (2015). Identification of a second GTP-bound magnesium ion in archaeal initiation factor 2. *Nucleic Acids Res.*, 43(5), 2946–2957.

Egea, P.F., Stroud, R.M., Walter, P. (2005). Targeting proteins to membranes: Structure of the signal recognition particle. *Curr. Opin. Struct. Biol.*, 15(2), 213–220.

Eichler, J. and Moll, R. (2001). The signal recognition particle of Archaea. *Trends Microbiol.*, 9(3), 130–136.

Erickson, F.L. and Hannig, E.M. (1996). Ligand interactions with eukaryotic translation initiation factor 2: Role of the gamma-subunit. *EMBO J.*, 15(22), 6311–6320.

Fan, H., Conn, A.B., Williams, P.B., Diggs, S., Hahm, J., Gamper, H.B., Jr., Hou, Y.M., O'Leary, S.E., Wang, Y., Blaha, G.M. (2017). Transcription-translation coupling: Direct interactions of RNA polymerase with ribosomes and ribosomal subunits. *Nucleic Acids Res.*, 45(19), 11043–11055.

Ferreira-Cerca, S. (2017). Life and death of ribosomes in Archaea. In *RNA Metabolism and Gene Expression in Archaea*, Clouet-d'Orval, B. (ed.). Springer, Cham.

Ferretti, M.B., Ghalei, H., Ward, E.A., Potts, E.L., Karbstein, K. (2017). Rps26 directs mRNA-specific translation by recognition of Kozak sequence elements. *Nat. Struct. Mol. Biol.*, 24, 700–707.

French, S.L., Santangelo, T.J., Beyer, A.L., Reeve, J.N. (2007). Transcription and translation are coupled in Archaea. *Mol. Biol. Evol.*, 24(4), 893–895.

Frolova, L., Le Goff, X., Rasmussen, H.H., Cheperegin, S., Drugeon, G., Kress, M., Arman, I., Haenni, A.L., Celis, J.E., Philippe, M. et al. (1994). A highly conserved eukaryotic protein family possessing properties of polypeptide chain release factor. *Nature*, 372(6507), 701–703.

Gabel, K., Schmitt, J., Schulz, S., Nather, D.J., Soppa, J. (2013). A comprehensive analysis of the importance of translation initiation factors for *Haloferax volcanii* applying deletion and conditional depletion mutants. *PLoS One*, 8(11), e77188.

Gerovac, M. and Tampe, R. (2019). Control of mRNA translation by versatile ATP-driven machines. *Trends Biochem. Sci.*, 44(2), 167–180.

Gouridis, G., Hetzert, B., Kiosze-Becker, K., de Boer, M., Heinemann, H., Nurenberg-Goloub, E., Cordes, T., Tampe, R. (2019). ABCE1 controls ribosome recycling by an asymmetric dynamic conformational equilibrium. *Cell Rep.*, 28(3), 723–734.

Graf, M., Huter, P., Maracci, C., Peterek, M., Rodnina, M.V., Wilson, D.N. (2018). Visualization of translation termination intermediates trapped by the apidaecin 137 peptide during RF3-mediated recycling of RF1. *Nat. Commun.*, 9(1), 3053.

Graille, M., Chaillet, M., van Tilbeurgh, H. (2008). Structure of yeast Dom34: A protein related to translation termination factor Erf1 and involved in No-Go decay. *J. Biol. Chem.*, 283(11), 7145–7154.

Grela, P., Bernado, P., Svergun, D., Kwiatowski, J., Abramczyk, D., Grankowski, N., Tchorzewski, M. (2008). Structural relationships among the ribosomal stalk proteins from the three domains of life. *J. Mol. Evol.*, 67(2), 154–167.

Grela, P., Szajwaj, M., Horbowicz-Drozdzal, P., Tchorzewski, M. (2019). How ricin damages the ribosome. *Toxins (Basel)*, 11(5), 241.

Gribaldo, S. and Cammarano, P. (1998). The root of the universal tree of life inferred from anciently duplicated genes encoding components of the protein-targeting machinery. *J. Mol. Evol.*, 47(5), 508–516.

Grill, S., Gualerzi, C.O., Londei, P., Blasi, U. (2000). Selective stimulation of translation of leaderless mRNA by initiation factor 2: Evolutionary implications for translation. *EMBO J.*, 19(15), 4101–4110.

de Groot, A., Roche, D., Fernandez, B., Ludanyi, M., Cruveiller, S., Pignol, D., Vallenet, D., Armengaud, J., Blanchard, L. (2014). RNA sequencing and proteogenomics reveal the importance of leaderless mRNAs in the radiation-tolerant bacterium *Deinococcus deserti*. *Genome Biol. Evol.*, 6(4), 932–948.

Grosjean, H., Gaspin, C., Marck, C., Decatur, W.A., de Crecy-Lagard, V. (2008). RNomics and modomics in the halophilic archaea *Haloferax volcanii*: Identification of RNA modification genes. *BMC Genomics*, 9, 470.

Guillon, J.M., Mechulam, Y., Blanquet, S., Fayat, G. (1993). Importance of formylability and anticodon stem sequence to give a tRNA(Met) an initiator identity in *Escherichia coli*. *J. Bacteriol.*, 175(14), 4507–4514.

Gupta, S., Roy, M., Ghosh, A. (2017). The archaeal signal recognition particle: Present understanding and future perspective. *Curr. Microbiol.*, 74(2), 284–297.

Harding, H.P., Zhang, Y., Zeng, H., Novoa, I., Lu, P.D., Calfon, M., Sadri, N., Yun, C., Popko, B., Paules, R. et al. (2003). An integrated stress response regulates amino acid metabolism and resistance to oxidative stress. *Mol. Cell.*, 11(3), 619–633.

Hartman, H. and Smith, T.F. (2010). GTPases and the origin of the ribosome. *Biol. Direct*, 5, 36.

Hasenöhrl, D., Benelli, D., Barbazza, A., Londei, P., Bläsi, U. (2006). *Sulfolobus solfataricus* translation initiation factor 1 stimulates translation initiation complex formation. *RNA*, 12(4), 674–682.

Hasenöhrl, D., Fabbretti, A., Londei, P., Gualerzi, C.O., Bläsi, U. (2009). Translation initiation complex formation in the crenarchaeon *Sulfolobus solfataricus*. *RNA*, 15(12), 2288–2298.

Hashimoto, T. and Hasegawa, M. (1996). Origin and early evolution of eukaryotes inferred from the amino acid sequences of translation elongation factors 1alpha/Tu and 2/G. *Adv. Biophys.*, 32, 73–120.

Hellen, C.U.T. (2018). Translation termination and ribosome recycling in eukaryotes. *Cold Spring Harb. Perspect. Biol.*, 10(10), a032656.

Heuer, A., Gerovac, M., Schmidt, C., Trowitzsch, S., Preis, A., Kotter, P., Berninghausen, O., Becker, T., Beckmann, R., Tampe, R. (2017). Structure of the 40S-ABCE1 post-splitting complex in ribosome recycling and translation initiation. *Nat. Struct. Mol. Biol.*, 24(1), 453–460.

Hinnebusch, A.G. (2017). Structural insights into the mechanism of scanning and start codon recognition in eukaryotic translation initiation. *Trends Biochem. Sci.*, 42(8), 589–611.

Huang, H.K., Yoon, H., Hannig, E.M., Donahue, T.F. (1997). GTP hydrolysis controls stringent selection of the AUG start codon during translation initiation in *Saccharomyces cerevisiae*. *Genes. Dev.*, 11(18), 2396–2413.

Huber, M., Faure, G., Laass, S., Kolbe, E., Seitz, K., Wehrheim, C., Wolf, Y.I., Koonin, E.V., Soppa, J. (2019). Translational coupling via termination-reinitiation in archaea and bacteria. *Nat. Commun.*, 10(1), 4006.

Ito, K., Honda, T., Suzuki, T., Miyoshi, T., Murakami, R., Yao, M., Uchiumi, T. (2014). Molecular insights into the interaction of the ribosomal stalk protein with elongation factor 1alpha. *Nucleic Acids Res.*, 42(22), 14042–14052.

Iwabe, N., Kuma, K., Hasegawa, M., Osawa, S., Miyata, T. (1989). Evolutionary relationship of archaebacteria, eubacteria, and eukaryotes inferred from phylogenetic trees of duplicated genes. *Proc. Natl. Acad. Sci. USA*, 86(23), 9355–9359.

Kapp, L.D. and Lorsch, J.R. (2004). GTP-dependent recognition of the methionine moiety on initiator tRNA by translation factor eIF2. *J. Mol. Biol.*, 335(4), 923–936.

Kazan, R., Bourgeois, G., Lazennec-Schurdevin, C., Larquet, E., Mechulam, Y., Coureux, P.D., Schmitt, E. (2022). Role of aIF5B in archaeal translation initiation. *Nucleic Acids Res.*, 50(11), 6532–6548.

Kim, K.K., Hung, L.W., Yokota, H., Kim, R., Kim, S.H. (1998). Crystal structures of eukaryotic translation initiation factor 5A from *Methanococcus jannaschii* at 1.8 A resolution. *Proc. Natl. Acad. Sci. USA*, 95(18), 10419–10424.

Kiosze-Becker, K., Ori, A., Gerovac, M., Heuer, A., Nurenberg-Goloub, E., Rashid, U.J., Becker, T., Beckmann, R., Beck, M., Tampe, R. (2016). Structure of the ribosome post-recycling complex probed by chemical cross-linking and mass spectrometry. *Nat. Commun.*, 7, 13248.

Knüppel, R., Trahan, C., Kern, M., Wagner, A., Grünberger, F., Hausner, W., Quax, T.E.F., Albers, S.V., Oeffinger, M. et al. (2021). Insights into synthesis and function of KsgA/Dim1-dependent rRNA modifications in archaea. *Nucleic Acids Res.*, 49(3), 1662–1687.

Kobayashi, K., Kikuno, I., Kuroha, K., Saito, K., Ito, K., Ishitani, R., Inada, T., Nureki, O. (2010). Structural basis for mRNA surveillance by archaeal Pelota and GTP-bound EF1α complex. *Proceedings of the National Academy of Sciences*, 107(41), 17575.

Kohler, R., Mooney, R.A., Mills, D.J., Landick, R., Cramer, P. (2017). Architecture of a transcribing-translating expressome. *Science*, 356(6334), 194–197.

Kyrpides, N.C. and Woese, C.R. (1998a). Archaeal translation initiation revisited: The initiation factor 2 and eukaryotic initiation factor 2B alpha-beta-delta subunit families. *Proc. Natl. Acad. Sci. USA*, 95(7), 3726–3730.

Kyrpides, N.C. and Woese C.R. (1998b). Universally conserved translation initiation factors. *Proc. Natl. Acad. Sci. USA*, 95(1), 224–228.

Lake, J.A., Henderson, E., Oakes, M., Clark, M.W. (1984). Eocytes: A new ribosome structure indicates a kingdom with a close relationship to eukaryotes. *Proc. Natl. Acad. Sci. USA*, 81(12), 3786–3790.

Lapointe, C.P., Grosely, R., Sokabe, M., Alvarado, C., Wang, J., Montabana, E., Villa, N., Shin, B.S., Dever, T.E., Fraser, C.S. et al. (2022). eIF5B and eIF1A reorient initiator tRNA to allow ribosomal subunit joining. *Nature*, 607, 185–190.

Lecompte, O., Ripp, R., Thierry, J.C., Moras, D., Poch, O. (2002). Comparative analysis of ribosomal proteins in complete genomes: An example of reductive evolution at the domain scale. *Nucleic Acids Res.*, 30(24), 5382–5390.

Leibundgut, M., Frick, C., Thanbichler, M., Böck, A., Ban, N. (2005). Selenocysteine tRNA-specific elongation factor SelB is a structural chimaera of elongation and initiation factors. *The EMBO Journal*, 24(1), 11–22.

Ling, C. and Ermolenko, D.N. (2016). Structural insights into ribosome translocation. *Wiley Interdiscip. Rev. RNA*, 7(5), 620–636.

Llacer, J.L., Hussain, T., Marler, L., Aitken, C.E., Thakur, A., Lorsch, J.R., Hinnebusch, A.G., Ramakrishnan, V. (2015). Conformational differences between open and closed states of the eukaryotic translation initiation complex. *Mol. Cell*, 59(3), 399–412.

Llacer, J.L., Hussain, T., Saini, A.K., Nanda, J.S., Kaur, S., Gordiyenko, Y., Kumar, R., Hinnebusch, A.G., Lorsch, J.R., Ramakrishnan, V. (2018). Translational initiation factor eIF5 replaces eIF1 on the 40S ribosomal subunit to promote start-codon recognition. *Elife*, 7, e39273.

Lo Gullo, G., Mattossovich, R., Perugino, G., La Teana, A., Londei, P., Benelli, D. (2019). Optimization of an in vitro transcription/translation system based on *Sulfolobus solfataricus* cell lysate. *Archaea*, 2019, 984253.

Lomakin, I.B., Kolupaeva, V.G., Marintchev, A., Wagner, G., Pestova, T.V. (2003). Position of eukaryotic initiation factor eIF1 on the 40S ribosomal subunit determined by directed hydroxyl radical probing. *Genes. Dev.*, 17(22), 2786–2797.

Londei, P., Altamura, S., Cammarano, P., Petrucci, L. (1986). Differential features of ribosomes and of poly(U)-programmed cell-free systems derived from sulphur-dependent archaebacterial species. *Eur. J. Biochem.*, 157(3), 455–462.

Lyu, Z. and Whitman, W.B. (2017). Evolution of the archaeal and mammalian information processing systems: Towards an archaeal model for human disease. *Cell Mol. Life Sci.*, 74(2), 183–212.

Maki, Y., Hashimoto, T., Zhou, M., Naganuma, T., Ohta, J., Nomura, T., Robinson, C.V., Uchiumi, T. (2007). Three binding sites for stalk protein dimers are generally present in ribosomes from archaeal organism. *J. Biol. Chem.*, 282(45), 32827–32833.

Martin, W. and Koonin, E.V. (2006). A positive definition of prokaryotes. *Nature*, 442(7105), 868–868.

Maruyama, K., Imai, H., Kawamura, M., Ishino, S., Ishino, Y., Ito, K., Uchiumi, T. (2019). Switch of the interactions between the ribosomal stalk and EF1A in the GTP- and GDP-bound conformations. *Sci. Rep.*, 9(15), 14761.

McGary, K. and Nudler, E. (2013). RNA polymerase and the ribosome: The close relationship. *Curr. Opin. Microbiol.*, 16(2), 112–117.

Mechulam, Y., Blanquet, S., Schmitt, E. (2011). Translation initiation. *EcoSal Plus*, 4, 2. doi: 10.1128/ecosalplus.4.2.2.

Melnikov, S., Mailliot, J., Rigger, L., Neuner, S., Shin, B.S., Yusupova, G., Dever, T.E., Micura, R., Yusupov, M. (2016). Molecular insights into protein synthesis with proline residues. *EMBO Rep.*, 17(12), 1776–1784.

Mitchell, S.F. and Lorsch, J.R. (2008). Should I stay or should I go? Eukaryotic translation initiation factors 1 and 1A control start codon recognition. *J. Biol. Chem.*, 283(41), 27345–27349.

Mochizuki, M., Kitamyo, M., Miyoshi, T., Ito, K., Uchiumi, T. (2012). Analysis of chimeric ribosomal stalk complexes from eukaryotic and bacterial sources: Structural features responsible for specificity of translation factors. *Genes Cells*, 17(4), 273–284.

Moll, I., Grill, S., Gualerzi, C.O., Blasi, U. (2002). Leaderless mRNAs in bacteria: Surprises in ribosomal recruitment and translational control. *Mol. Microbiol.*, 43(1), 239–246.

Moll, I., Hirokawa, G., Kiel, M.C., Kaji, A., Blasi, U. (2004). Translation initiation with 70S ribosomes: An alternative pathway for leaderless mRNAs. *Nucleic Acids Res.*, 32(11), 3354–3363.

Monestier, A., Lazennec-Schurdevin, C., Coureux, P.D., Mechulam, Y., Schmitt, E. (2018). Role of aIF1 in Pyrococcus abyssi translation initiation. *Nucleic Acids Res.*, 46(20), 11061–11074.

Naganuma, T., Nomura, N., Yao, M., Mochizuki, M., Uchiumi, T., Tanaka, I. (2010). Structural basis for translation factor recruitment to the eukaryotic/archaeal ribosomes. *J. Biol. Chem.*, 285(7), 4747–4756.

Narrowe, A.B., Spang, A., Stairs, C.W., Caceres, E.F., Baker, B.J., Miller, C.S., Ettema, T.J.G. (2018). Complex evolutionary history of translation elongation factor 2 and diphthamide biosynthesis in Archaea and parabasalids. *Genome Biol. Evol.*, 10(9), 2380–2393.

Naveau, M., Lazennec-Schurdevin, C., Panvert, M., Dubiez, E., Mechulam, Y., Schmitt, E. (2013). Roles of yeast eIF2alpha and eIF2beta subunits in the binding of the initiator methionyl-tRNA. *Nucleic Acids Res.*, 41(9), 1047–1057.

Nissen, P., Kjeldgaard, M., Thirup, S., Polekhina, G., Reshetnikova, L., Clark, B.F.C., Nyborg, J. (1995). Crystal structure of the ternary complex of Phe-$tRNA^{Phe}$, EF-Tu, and a GTP analog. *Science*, 270(5241), 1464–1472.

Nomura, N., Honda, T., Baba, K., Naganuma, T., Tanzawa, T., Arisaka, F., Noda, M., Uchiyama, S., Tanaka, I., Yao, M. et al. (2012). Archaeal ribosomal stalk protein interacts with translation factors in a nucleotide-independent manner via its conserved C terminus. *Proc. Natl. Acad. Sci. USA*, 109(10), 3748–3753.

Nurenberg-Goloub, E., Kratzat, H., Heinemann, H., Heuer, A., Kotter, P., Berninghausen, O., Becker, T., Tampe, R., Beckmann, R. (2020). Molecular analysis of the ribosome recycling factor ABCE1 bound to the 30S post-splitting complex. *EMBO J.*, 39(9), e103788. doi: 10.15252/embj.2019103788.

Ortiz, P.A., Ulloque, R., Kihara, G.K., Zheng, H., Kinzy, T.G. (2006). Translation elongation factor 2 anticodon mimicry domain mutants affect fidelity and diphtheria toxin resistance. *J. Biol. Chem.*, 281(43), 32639–32648.

Pace, N.R. (2006). Time for a change. *Nature*, 441, 289–289.

Pakos-Zebrucka, K., Koryga, I., Mnich, K., Ljujic, M., Samali, A., Gorman, A.M. (2016). The integrated stress response. *EMBO Rep.*, 17, 1374–1395.

Park, M.H., Nishimura, K., Zanelli, C.F., Valentini, S.R. (2010). Functional significance of eIF5A and its hypusine modification in eukaryotes. *Amino Acids*, 38(2), 491–500.

Peat, T.S., Newman, J., Waldo, G.S., Berendzen, J., Terwilliger, T.C. (1998). Structure of translation initiation factor 5A from Pyrobaculum aerophilum at 1.75 A resolution. *Structure*, 6(9), 1207–1214.

Pedulla, N., Palermo, R., Hasenohrl, D., Blasi, U., Cammarano, P., Londei, P. (2005). The archaeal eIF2 homologue: Functional properties of an ancient translation initiation factor. *Nucleic Acids Res.*, 33(6), 1804–1812.

Perrochia, L., Crozat, E., Hecker, A., Zhang, W., Bareille, J., Collinet, B., van Tilbeurgh, H., Forterre, P., Basta, T. (2013). In vitro biosynthesis of a universal t6A tRNA modification in Archaea and Eukarya. *Nucleic Acids Res.*, 41(3), 1953–1964.

Pestova, T.V., Lomakin, I.B., Lee, J.H., Choi, S.K., Dever, T.E., Hellen, C.U. (2000). The joining of ribosomal subunits in eukaryotes requires eIF5B. *Nature*, 403(13), 332–335.

Pestova, T.V., de Breyne, S., Pisarev, A.V., Abaeva, I.S., Hellen, C.U. (2008). eIF2-dependent and eIF2-independent modes of initiation on the CSFV IRES: A common role of domain II. *EMBO J.*, 27, 1060–1072.

Pierson, W.E., Hoffer, E.D., Keedy, H.E., Simms, C.L., Dunham, C.M., Zaher, H.S. (2016). Uniformity of peptide release is maintained by methylation of release factors. *Cell Rep*, 17(1), 11–18.

Pisarev, A.V., Skabkin, M.A., Pisareva, V.P., Skabkina, O.V., Rakotondrafara, A.M., Hentze, M.W., Hellen, C.U., Pestova, T.V. (2010). The role of ABCE1 in eukaryotic posttermination ribosomal recycling. *Mol. Cell*, 37(2), 196–210.

Prunetti, L., Graf, M., Blaby, I.K., Peil, L., Makkay, A.M., Starosta, A.L., Papke, R.T., Oshima, T., Wilson, D.N., de Crecy-Lagard, V. (2016). Deciphering the translation initiation factor 5A modification pathway in halophilic Archaea. *Archaea*, 2016, 7316725.

Reibarkh, M., Yamamoto, Y., Singh, C.R., del Rio, F., Fahmy, A., Lee, B., Luna, R.E., Ii M, Wagner, G., Asano, K. (2008). Eukaryotic initiation factor (eIF) 1 carries two distinct eIF5-binding faces important for multifactor assembly and AUG selection. *J. Biol. Chem.*, 283(2), 1094–1103.

Rivera, M.C. and Lake, J.A. (1992). Evidence that eukaryotes and eocyte prokaryotes are immediate relatives. *Science*, 257(5066), 74–76.

Rodnina, M.V. (2018). Translation in prokaryotes. *Cold Spring Harb Perspect Biol.*, 10(9), a032664.

Roger, A.J., Sandblom, O., Doolittle, W.F., Philippe, H. (1999). An evaluation of elongation factor 1 alpha as a phylogenetic marker for eukaryotes. *Mol. Biol. Evol.*, 16(2), 218–233.

Roll-Mecak, A., Alone, P., Cao, C., Dever, T.E., Burley, S.K. (2004). X-ray structure of translation initiation factor eIF2gamma: Implications for tRNA and eIF2alpha binding. *J. Biol. Chem.*, 279(11), 10634–10642.

Romero, H., Zhang, Y., Gladyshev, V.N., Salinas, G. (2005). Evolution of selenium utilization traits. *Genome Biol.*, 6(8), R66.

Rother, M., Wilting, R., Commans, S., Böck, A. (2000). Identification and characterisation of the selenocysteine-specific translation factor SelB from the archaeon *Methanococcus jannaschii*. *J. Mol. Biol.*, 299(2), 351–358.

Saini, A.K., Nanda, J.S., Lorsch, J.R., Hinnebusch, A.G. (2010). Regulatory elements in eIF1A control the fidelity of start codon selection by modulating tRNA(i)(Met) binding to the ribosome. *Genes Dev.*, 24(1), 97–110.

Saito, K., Kobayashi, K., Wada, M., Kikuno, I., Takusagawa, A., Mochizuki, M., Uchiumi, T., Ishitani, R., Nureki, O., Ito, K. (2010). Omnipotent role of archaeal elongation factor 1 alpha (EF1alpha in translational elongation and termination, and quality control of protein synthesis. *Proc. Natl. Acad. Sci. USA*, 107(45), 19242–19247.

Sas-Chen, A., Thomas, J.M., Matzov, D., Taoka, M., Nance, K.D., Nir, R., Bryson, K.M., Shachar, R., Liman, G.L.S., Burkhart, B.W. et al. (2020). Dynamic RNA acetylation revealed by quantitative cross-evolutionary mapping. *Nature*, 583. doi: 10.1038/s41586-020-2418-2.

Sawyer, E.B. and Cortes, T. (2022). Ribosome profiling enhances understanding of mycobacterial translation. *Front Microbiol.*, 13, 976550.

Schmitt, E., Blanquet, S., Mechulam, Y. (2002). The large subunit of initiation factor aIF2 is a close structural homologue of elongation factors. *EMBO J.*, 21(7), 1821–1832.

Schmitt, E., Naveau, M., Mechulam, Y. (2010). Eukaryotic and archaeal translation initiation factor 2: A heterotrimeric tRNA carrier. *FEBS Lett.*, 584(2), 405–412.

Schmitt, E., Panvert, M., Lazennec-Schurdevin, C., Coureux, P.D., Perez, J., Thompson, A., Mechulam, Y. (2012). Structure of the ternary initiation complex aIF2-GDPNP-methionylated initiator tRNA. *Nat. Struct. Mol. Biol.*, 19, 450–454.

Schmitt, E., Coureux, P.D., Monestier, A., Dubiez, E., Mechulam, Y. (2019). Start codon recognition in eukaryotic and archaeal translation initiation: A common structural core. *Int. J. Mol. Sci.*, 20(4), 939.

Shao, S., Murray, J., Brown, A., Taunton, J., Ramakrishnan, V., Hegde, R.S. (2016). Decoding mammalian ribosome-mRNA states by translational GTPase complexes. *Cell*, 167(5), 1229–1240.

Shatsky, I.N., Terenin, I.M., Smirnova, V.V., Andreev, D.E. (2018). Cap-independent translation: What's in a name? *Trends Biochem. Sci.*, 43(11), 882–895.

Shin, B.S., Kim, J.R., Walker, S.E., Dong, J., Lorsch, J.R., Dever, T.E. (2011). Initiation factor eIF2gamma promotes eIF2-GTP-Met-tRNAi(Met) ternary complex binding to the 40S ribosome. *Nat. Struct. Mol. Biol.*, 18(11), 1227–1234.

Shoemaker, C.J. and Green, R. (2011). Kinetic analysis reveals the ordered coupling of translation termination and ribosome recycling in yeast. *Proc. Natl. Acad. Sci. USA*, 108(51), E1392–E1398.

Shoemaker, C.J. and Green, R. (2012). Translation drives mRNA quality control. *Nat. Struct. Mol. Biol.*, 19, 594–601.

Shoemaker, C.J., Eyler, D.E., Green, R. (2010). Dom34:Hbs1 promotes subunit dissociation and peptidyl-tRNA drop-off to initiate no-go decay. *Science*, 330(6002), 369–372.

Sievers, A., Beringer, M., Rodnina, M.V., Wolfenden, R. (2004). The ribosome as an entropy trap. *Proc. Natl. Acad. Sci. USA*, 101(21), 7897–7901.

Sokabe, M., Yao, M., Sakai, N., Toya, S., Tanaka, I. (2006). Structure of archaeal translational initiation factor 2 betagamma-GDP reveals significant conformational change of the beta-subunit and switch 1 region. *Proc. Natl. Acad. Sci. USA*, 103(35), 13016–13021.

Srinivasan, G., James, C.M., Krzycki, J.A. (2002). Pyrrolysine encoded by UAG in Archaea: Charging of a UAG-decoding specialized tRNA. *Science*, 296(5572), 1459–1462.

Stolboushkina, E., Nikonov, S., Nikulin, A., Blasi, U., Manstein, D.J., Fedorov, R., Garber, M., Nikonov, O. (2008). Crystal structure of the intact archaeal translation initiation factor 2 demonstrates very high conformational flexibility in the alpha- and beta-subunits. *J. Mol. Biol.*, 382(3), 680–691.

Su, X., Lin, Z., Lin, H. (2013). The biosynthesis and biological function of diphthamide. *Crit. Rev. Biochem. Mol. Biol.*, 48(6), 515–521.

Suzuki, T., Ito, K., Miyoshi, T., Murakami, R., Uchiumi, T. (2021). Structural insights into the switching off of the interaction between the archaeal ribosomal stalk and aEF1A by nucleotide exchange factor aEF1B. *Journal of Molecular Biology*, 433(15), 167046.

Tanzawa, T., Kato, K., Girodat, D., Ose, T., Kumakura, Y., Wieden, H.J., Uchiumi, T., Tanaka, I., Yao, M. (2018). The C-terminal helix of ribosomal P stalk recognizes a hydrophobic groove of elongation factor 2 in a novel fashion. *Nucleic Acids Res.*, 46(6), 3232–3244.

Terenin, I.M., Dmitriev, S.E., Andreev, D.E., Shatsky, I.N. (2008). Eukaryotic translation initiation machinery can operate in a bacterial-like mode without eIF2. *Nat. Struct. Mol. Biol.*, 15(8), 836–841.

Thakur, A. and Hinnebusch, A.G. (2018). eIF1 Loop 2 interactions with Met-tRNAi control the accuracy of start codon selection by the scanning preinitiation complex. *Proc. Natl. Acad. Sci. USA*, 115(18), E4159–E4168.

Toffano-Nioche, C., Ott, A., Crozat, E., Nguyen, A.N., Zytnicki, M., Leclerc, F., Forterre, P., Bouloc, P., Gautheret, D. (2013). RNA at 92 degrees C: The non-coding transcriptome of the hyperthermophilic archaeon *Pyrococcus abyssi*. *RNA Biol.*, 10(7), 1211–1220.

Torarinsson, E., Klenk, H.P., Garrett, R.A. (2005). Divergent transcriptional and translational signals in Archaea. *Environ Microbiol*, 7(1), 47–54.

Tsuboi, T., Kuroha, K., Kudo, K., Makino, S., Inoue, E., Kashima, I., Inada, T. (2012). Dom34:hbs1 plays a general role in quality-control systems by dissociation of a stalled ribosome at the 3' end of aberrant mRNA. *Mol. Cell*, 46(4), 518–529.

Vetter, I.R. and Wittinghofer, A. (2001). The guanine nucleotide-binding switch in three dimensions. *Science*, 294(5545), 1299–1304.

Vitagliano, L., Masullo, M., Sica, F., Zagari, A., Bocchini, V. (2001). The crystal structure of Sulfolobus solfataricus elongation factor 1alpha in complex with GDP reveals novel features in nucleotide binding and exchange. *EMBO J.*, 20(19), 5305–5311.

Voorhees, R.M., Schmeing, T.M., Kelley, A.C., Ramakrishnan, V. (2010). The mechanism for activation of GTP hydrolysis on the ribosome. *Science*, 330(6005), 835–838.

Weisser, M., Voigts-Hoffmann, F., Rabl, J., Leibundgut, M., Ban, N. (2013). The crystal structure of the eukaryotic 40S ribosomal subunit in complex with eIF1 and eIF1A. *Nat. Struct. Mol. Biol.*, 20(8), 1015–1017.

Woese, C.R. and Fox, G.E. (1977). Phylogenetic structure of the prokaryotic domain: The primary kingdoms. *Proc. Natl. Acad. Sci. USA*, 74(11), 5088–5090.

Woese, C.R., Gutell, R., Gupta, R., Noller, H.F. (1983). Detailed analysis of the higher-order structure of 16S-like ribosomal ribonucleic acids. *Microbiol. Rev.*, 47(4), 621–669.

Woese, C.R., Kandler, O., Wheelis, M.L. (1990). Towards a natural system of organisms: Proposal for the domains Archaea, Bacteria, and Eucarya. *Proc. Natl. Acad. Sci. USA*, 87(12), 4576–4579.

Yanagisawa, T., Ishii, R., Hikida, Y., Fukunaga, R., Sengoku, T., Sekine, S., Yokoyama, S. (2015). A SelB/EF-Tu/aIF2γ-like protein from *Methanosarcina mazei* in the GTP-bound form binds cysteinyl-tRNA(Cys.). *J. Struct. Funct. Genomics*, 16(1), 25–41.

Yanofsky, C. (1981). Attenuation in the control of expression of bacterial operons. *Nature*, 289, 751–758.

Yao, M., Ohsawa, A., Kikukawa, S., Tanaka, I., Kimura, M. (2003). Crystal structure of hyperthermophilic archaeal initiation factor 5A: A homologue of eukaryotic initiation factor 5A (eIF-5A). *J. Biochem.*, 133(1), 75–81.

Yatime, L., Schmitt, E., Blanquet, S., Mechulam, Y. (2004). Functional molecular mapping of archaeal translation initiation factor 2. *J. Biol. Chem.*, 279(16), 15984–15993.

Yatime, L., Mechulam, Y., Blanquet, S., Schmitt, E. (2006). Structural switch of the gamma subunit in an archaeal aIF2 alpha gamma heterodimer. *Structure*, 14(1), 119–128.

Yutin, N., Puigbo, P., Koonin, E.V., Wolf, Y.I. (2012). Phylogenomics of prokaryotic ribosomal proteins. *PLoS One*, 7(5), e36972.

Zavialov, A.V., Buckingham, R.H., Ehrenberg, M. (2001). A posttermination ribosomal complex is the guanine nucleotide exchange factor for peptide release factor RF3. *Cell*, 107(1), 115–124.

Zwieb, C. and Bhuiyan, S. (2010). Archaea signal recognition particle shows the way. *Archaea*, 485051. doi: 10.1155/2010/485051.

List of Authors

Alexey ALEKSANDROV
Laboratoire d'optique et biosciences
Inserm, CNRS
Institut Polytechnique de Paris
Palaiseau
France

Tamara BASTA
Institute for Integrative Biology of the Cell (I2BC)
CEA, CNRS
Université Paris-Saclay
Gif-sur-Yvette
France

Manon BATISTA
MCD-CBI, CNRS
Université de Toulouse
France

Anaïs BAYARD
Laboratoire d'optique et biosciences
Inserm, CNRS
Institut Polytechnique de Paris
Palaiseau
France

Hubert F. BECKER
Laboratoire d'optique et biosciences
Inserm, CNRS
Institut Polytechnique de Paris
Palaiseau
and
Sorbonne Université
Paris
France

Marie BOUVIER
MCD-CBI, CNRS
Université de Toulouse
France

Béatrice CLOUET-D'ORVAL
MCD-CBI, CNRS
Université de Toulouse
France

Sébastien FERREIRA-CERCA
Cellular Biochemistry of Microorganisms
Institute for Biochemistry,
Genetics and Microbiology
Regensburg Center for Biochemistry
University of Regensburg
Germany
and
BIOC, CNRS
Institut Polytechnique de Paris
Palaiseau
France

Didier FLAMENT
BEEP, CNRS
Ifremer
Université de Bretagne Occidentale
Plouzané
France

Thomas FOUQUEAU
RNAP
ISMB
University of London (UCL)
United Kingdom

Bruno FRANZETTI
IBS, CNRS
Université Grenoble Alpes
France

Christine GASPIN
MIAT, INRAE
Université de Toulouse
Castanet-Tolosan
France

Ghislaine HENNEKE
BEEP, CNRS
Ifremer
Université de Bretagne Occidentale
Plouzané
France

Duy KHANH PHUNG
RNAP
ISMB
University of London (UCL)
United Kingdom

Caroline L'HERMITTE-STEAD
Laboratoire d'optique et biosciences
Inserm, CNRS
Institut Polytechnique de Paris
Palaiseau
France

Roxane LESTINI
Laboratoire d'optique et biosciences
Inserm, CNRS
Institut Polytechnique de Paris
Palaiseau
France

Yves MECHULAM
BIOC, CNRS
Institut Polytechnique de Paris
Palaiseau
France

Hannu MYLLYKALLIO
Laboratoire d'optique et biosciences
Inserm, CNRS
Institut Polytechnique de Paris
Palaiseau
France

Marc NADAL
IBENS, CNRS
Inserm
PSL University
Paris
France

Philippe OGER
MAP, CNRS
INSA Lyon
Villeurbanne
France

Ludovic SAUGUET
Institut Pasteur
Université Paris Cité
France

Emmanuelle SCHMITT
BIOC, CNRS
Institut Polytechnique de Paris
Palaiseau
France

Index

G, H, I

L, M

N, O

P

R

S

T, X